Algorithms for Intelligent Systems

This book series publishes research on the analysis and development of algorithms for intelligent systems with their applications to various real world problems. It covers research related to autonomous agents, multi-agent systems, behavioral modeling, reinforcement learning, game theory, mechanism design, machine learning, meta-heuristic search, optimization, planning and scheduling, artificial neural networks, evolutionary computation, swarm intelligence and other algorithms for intelligent systems.

The book series includes recent advancements, modification and applications of the artificial neural networks, evolutionary computation, swarm intelligence, artificial immune systems, fuzzy system, autonomous and multi agent systems, machine learning and other intelligent systems related areas. The material will be beneficial for the graduate students, post-graduate students as well as the researchers who want a broader view of advances in algorithms for intelligent systems. The contents will also be useful to the researchers from other fields who have no knowledge of the power of intelligent systems, e.g. the researchers in the field of bioinformatics, biochemists, mechanical and chemical engineers, economists, musicians and medical practitioners.

The series publishes monographs, edited volumes, advanced textbooks and selected proceedings.

More information about this series at http://www.springer.com/series/16171

G. S. Tomar · Ashish Bagwari

Fundamentals of Electronic Devices and Circuits

Springer

G. S. Tomar
Machine Intelligence Research Labs
Gwalior, Madhya Pradesh, India

T.H.D.C Institute of Hydropower
Engineering and Technology
Tehri Garhwal, Uttarakhand, India

Ashish Bagwari
Department of Electronics and
Communication Engineering
Women Institute of Technology
Dehradun, Uttarakhand, India

Uttarakhand Technical University
Dehradun, Uttarakhand, India

ISSN 2524-7565 ISSN 2524-7573 (electronic)
Algorithms for Intelligent Systems
ISBN 978-981-15-0269-9 ISBN 978-981-15-0267-5 (eBook)
https://doi.org/10.1007/978-981-15-0267-5

This Springer imprint is published by the registered company Springer Nature Singapore Pte Ltd.
The registered company address is: 152 Beach Road, #21-01/04 Gateway East, Singapore 189721, Singapore

Dedicated to my real God my Parents "Shri Pradeep Bagwari and Shrimati Saraswati Bagwari," Baba Kedarnath, my better half, Anshul, Anuskha and Agrim.

Dr. Ashish Bagwari

Preface

In recent years, electronic systems play an important role in human life. The importance of electronics is well known in various fields of engineering. It is therefore necessary for an electronic and electrical engineer to know the fundamentals of electronic system or network.

The overwhelming response to our books on Fundamental of Electronic Devices and Circuits, written according to the syllabus of electronics systems offered by various Indian universities as well as universities abroad, inspired us to write this book.

This book has been prepared carefully, and the background of each topic is given; suitable practical illustrations, numerical examples, and detailed explanation of each step given make it easy for students to understand the complicated derivations.

In every section of each chapter, the important basic concepts are highlighted by Key Points, which is the feature of this book. A large number of solved problems are given in each chapter, which help the students to understand the theory in great depth. The whole book has been divided into seven chapters:

Chapter 1 explains crystal properties and charge carriers in semiconductors and magnetic material.
Chapter 2 explains special diodes and linear wave shaping.
Chapter 3 discusses transistor amplifier frequency response.
Chapter 4 introduces feedback system.
Chapter 5 introduces oscillators.
Chapter 6 discusses OP-AMP applications, timer, voltage regulator, and converter.
Chapter 7 explains multistage and tuned amplifiers.

The authors hope that this book will prove to be very useful not only to the students but also to the subject faculties. The students have to omit nothing and possibly have to cover nothing more, and we welcome any suggestions toward the improvement of this book.

Tehri Garhwal, India
Dehradun, India

Prof. (Dr.) G. S. Tomar
Dr. Ashish Bagwari

Acknowledgements

We express our deep gratitude to all those who helped in making this book a reality. Much needed moral support and encouragement are provided on numerous occasions by our whole family.

We are very thankful to Prof. Kim Mallalu, Department of Electrical and Computer Engineering, University of the West Indies, for his time to time, much-needed, valuable guidance. Without the full support and cheerful encouragement of Dr. Ashraf Samarah, Department of Electrical Energy Engineering, Al-Balqa' Applied University, Jordan, and Dr. Alhad Kuwadekar, University of South Wales, Cardiff, UK, this book would not have been completed on time.

I (Dr. Bagwari) thank my parents Shri Pradeep Bagwari and Shrimati Saraswati Bagwari, who had been encouraging and supporting us throughout the tenure of this work, and Grace of Baba Kedarnath to achieve this goal. I also wish to acknowledge my better half Shrimati Jyotshana Bagwari, family members, WIT faculties, staff members, and WIT students for their continuous support, help, and encouragement without which this book would not have been completed on time. Last but not least, special thanks to my lovely nephew Mr. Anshul Semwal, niece Miss. Anushkha Gaur, and my son Mr. Agrim Bagwari for their true love.

Any suggestions for the improvement of this book will be acknowledged and appreciated.

Prof. (Dr.) G. S. Tomar
Dr. Ashish Bagwari

About This Book

This book has a broad concept and basic conceptual inputs. This book contains seven chapters, and each and every chapter carries suitable block diagrams, circuit diagram, easy and understandable language, valuable and interesting solved examples, and important test questions as well.

This book is suitable for various Indian universities as well as universities abroad and also targeting the basic concepts, which are at par with the inputs needed for competitive exams and GATE, etc. This book is helpful for engineering groups like B.Tech./B.E./B.S., M.Tech./M.E./M.S. (electronics and communication and electrical engineering). Both the students and teachers will find this book stimulating as it contains around 100+ well-annotated examples. It also includes 75+ numerical problems to test the students' understanding of the concepts discussed.

In competition point of view, it will be very relevant because it contains worked examples, case studies, solutions to problems, and others as well, which are beneficial to the readers.

It is suitable for professionals related to electronics/electrical background. It will be helpful to understand the basics of electronic circuitry, basic concepts of feedback, amplifier, oscillator, applications of OP-AMP, and other relevant topics as well.

- This book offers an introductory study of the conceptual development of the subject.
- It provides a simple and lucid presentation of the essential principles, formulae, and definitions of electronic devices and circuits.
- It presents several new illustrations, examples, and exercises, and further readings have been specially designed in the text throughout.
- It gives a large number of worked-out examples to illustrate the theory and to demonstrate their use in practical designs.
- It provides short questions with answers at the end of each chapter.

Contents

About the Authors

Dr. G. S. Tomar holds UG, PG, and Ph.D. degrees in Electronics Engineering from Institute of Engineering Calcutta, MNREC Allahabad and RGPV Bhopal respectively and a PDF from the University of Kent, UK. He is the Director of THDC Hydropower engineering and Technology, Tehri Garhwal, India. Has served at the Department of Electrical Computer Engineering, University of West Indies, Trinidad and Tobago, and the University of Kent, Canterbury, UK, MITS, SRCEM and IIITM Gwalior. His research areas include air-interface and advanced communication networks, sensors and sensor networks, and digital design. He is actively involved in IEEE activities and has organized more than 30 IEEE international conferences in India and other countries. He has delivered keynote at various international conferences. He is associated with professional societies like IEEE, and is a senior member of the IEEE, and Fellow IETE and IE (I). He is a Visiting Professor at Hannam University, Korea; and several other reputed institutes. He is the chief editor of 5 international journals and has published more than 190 research papers in international journals/conferences and has authored 11 books and 10 book chapters.

Dr. Ashish Bagwari holds B.Tech. (with Honors), M.Tech. (Gold Medalist), and Ph.D. degrees in Electronics and Communication Engineering. He is currently Head of the Electronics and Communication Engineering Department, Women's Institute of Technology (WIT) (Institute of State Government), Dehradun, affiliated to Uttarakhand Technical University (State Government Technical University), Dehradun, India. He also has more than 9 years' experience in industry, academic, and research. Dr. Bagwari had associated with National Institute of Technology, Kurukshetra as a Research Scholar, Reliance Pvt. Ltd. Mumbai as an Executive Engineer and visited China also. He received the Best WIT Faculty Award in 2013 and 2015 and Best Project Guide Award in 2015. Dr. Bagwari also received the Corps of Electrical and Mechanical Engineers Prize from the Institution of Engineers, India (IEI), in December 2015 and was named in Who's Who in the World 2016 and 2017. Dr. Bagwari has published more than 95 research papers in various international journals (including ISI/SCI indexed) and IEEE international conferences. His current research interests include wireless communications, sensor networks, 5-G technology, digital communication, and mobile ad hoc networks. He is a Senior Member of IEEE USA, MIR Laboratories India, IETE, ACM, and IAENG. He has also been an editor, an advisor, and a reviewer of several well-known international journals published by IEEE, Taylor & Francis, Springer, Elsevier, IJCCN, ISTP, IJATER, JREEE, JCSR, and JNMS, CICN-2011, 2013, 2015, 2016, 2017, CSNT-2014, 2015, 2016, 2017, ICMWOC-2014, I4CT'2014, and ICEPIT-2014. Dr. Bagwari has organized and been a key speaker at several conferences/workshops. Dr. Bagwari has filed two patents, and has written four books and more than 10 book chapters/articles.

Chapter 1
Crystal Properties and Charge Carriers in Semiconductors and Magnetic Material

Learning Objectives

- **Define semiconductor material, various bands like conduction band, valence band, and energy band gap.**
- **Discuss classification of semiconductor materials, such as element material and compound material, and their types.**
- **Introduce crystal lattice with their classifications.**
- **Discuss types of solids.**
- **Discuss various types of magnetization.**

1.1 Introduction

Basically, crystal is a solid in which regular pattern repeats itself, while semiconductors are those materials whose conductivities are greater than insulator but lesser than conductors.

In this chapter, we discuss the semiconductor introduction, energy bands, types of crystals, and classifications of magnetization.

1.1.1 Semiconductor

Semiconductors are those materials whose conductivities exist between conductors and insulators. They have poor conductivity than conductors and higher than insulators. When the temperature of a semiconductor is increased, its resistivity decreases or conductivity increases; therefore, the semiconductor has negative temperature

G. S. Tomar and A. Bagwari, *Fundamentals of Electronic Devices and Circuits*, Algorithms for Intelligent Systems, https://doi.org/10.1007/978-981-15-0267-5_1

coefficient of resistance. It is significant that the conductivity of these materials can be varied over orders of magnitude by changes in temperature optical excitation and impurity content.

There are different types of semiconductor materials such as gallium nitride (GaN), gallium phosphide (GaP), gallium aronide (GaAs), zinc sulfide (ZnS), and indium antimonide (InSb).

1.1.2 Energy Band

The range of energies possessed by electrons of the same orbit in a solid is called energy band.

1.1.3 Valence Band

The energy band which possesses the valence electrons is called valence band. This band may be partially or completely filled with electrons, i.e., band does not remain empty.

1.1.4 Conduction Band

The energy band which possesses the free electrons is called conduction band. An electron in this band takes place in conducting region.

1.1.5 Energy Band Gap

Between the valence band and the conduction band, there is an energy gap "Eg." This is called energy band gap or forbidden energy gap. Electrons are not found in this band for semiconductors: $Eg < 3$ eV; for germanium: $Eg = 0.72$ eV; and for silicon: $Eg = 1.12$ eV.

1.2 Elemental and Compound Semiconductor Materials

Elemental semiconductors are single-element semiconductors. They belong to the IV group of the periodic table, e.g., Si, Ge, C, Sn.

Compound semiconductors are formed by using two or more elements. They are synthesized using elements from group II to group VI of the periodic table, e.g., from group III and group V (III–V compounds) or II and VI (II–VI compounds), **e.g.,** III–V semiconductors: GaAs, GaP, GaN, GaAlAs, InP, InSb, etc., II–VI semiconductors: CdSe, CdTe, CdHgTe, and ZnS. A compound semiconductor elements can also form binary (two elements, e.g., gallium (III) arsenide (GaAs)], ternary [three elements, e.g., indium gallium arsenide (InGaAs)] and quaternary [four elements, e.g., aluminum gallium indium phosphide (AlInGaP)] alloys.

1.2.1 Semiconductor Material

The magic word semiconductor is composed of two words—semi and conductor. Semi means not complete, while conductor means something which can conduct electricity. In other words, semiconductors are a group of material has electrical conductivity which lies between or exits between metal (conductor and insulator). The conductivity of material can be varied over the order of magnitude by changing temperature and adding impurity.

1.2.2 Classification of Semiconductor Materials

There are two types of semiconductor materials—one is elemental material and other is compound material as shown in Fig. 1.1.

1.2.2.1 Elemental Material

Elemental material contains single element which means it does not have any combination of elements and these are available in natural form.

Example: Silicon (Si), Germanium (Ge).

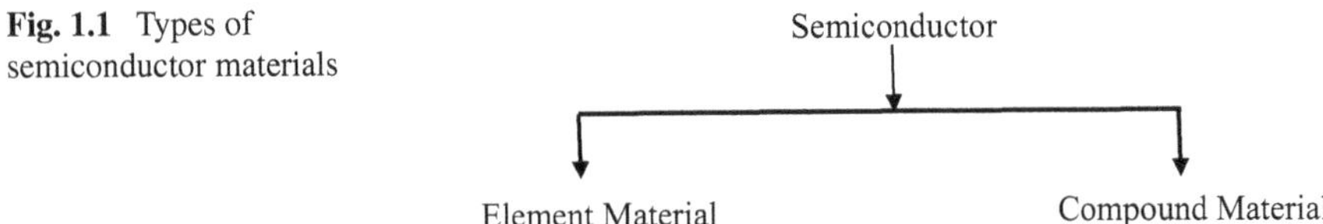

Fig. 1.1 Types of semiconductor materials

1.2.2.2 Compound Material

This kind of material contains more than one element material or we can say that it is having composed element to make a compound material.

Example: GaN, Gap, GaAs, InP, ZnS, InSb, Cds, Sic, SiGe.

1.2.3 Types of Compound Materials

From the types of compound materials, we further dividing compound materials into three categories:

1.2.3.1 Binary compound material.
1.2.3.2 Ternary compound material.
1.2.3.3 Quaternary compound material.

1.2.3.1 Binary Compound Material: These materials are the combination or composition of two elements. Gallium arsenide (GaAs) is an example of such kind of compound material.

1.2.3.2 Ternary Compound Material: These materials are the combination or composition of three elemental materials. Gallium arsenide phosphide (GaAsP) is an example of such kind of compound material.

1.2.3.3 Quaternary Compound Material: These materials are the combination of four elemental materials. Example is indium gallium arsenide phosphide (InGaAsP).

1.3 Crystal Lattice Structure

Crystal is solid in which regular pattern is repeating itself. The atoms in a crystal are in a regular repeating pattern called the crystalline lattice. By using scanning tunneling microscope (STM), individual can be observed. Graphite and diamond are made of same carbon atom. The difference is how those atoms are arranged, and to describe crystal structure, we must distinguish between the patterns of repetition.

1.3.1 Crystal Lattice

The crystal lattice is an arrangement of atoms in various solid. The atoms are combination of electron, proton, and neutron. The proton and neutron makeup the center of the atom called nucleus, and the electron flies around above the nucleus in a small cloud. The electron carries the negative charge particles, and protons carry positive charge carrier. In normal atom, the no of electrons and protons is equal. The negative electrons are attached to positive nucleus by the same electric force which causes magnets to work that is why atoms are together.

1.3.1.1 Classifications of Crystal Lattice

Crystal lattice can be classified into seven ways.

1.3.1.1.1 Cubic lattice structure: This is a simple cubic system that has one lattice point on each corner of the cube with each lattice point shared equal between eight adjacent cubes.

(i) In body-centered cubic system, there are eight corner points and one lattice point in the center of the unit cell.
(ii) In face-centered cubic system, lattice points are on the face of the cube.

1.3.1.1.2 Tetragonal lattice structure: It is similar to cubic crystals but longer along one axis than the other as shown in Fig. 1.2.

1.3.1.1.3 Orthorhombic lattice structure: It likes tetragonal crystal, except their cross section is not square. They form rhombic prism or di-pyramids (two pyramids stucked together) as shown in Fig. 1.3.

1.3.1.1.4 Hexagonal cubic lattice structure: Figure 1.4 shows the structure of hexagonal cubic, there is six-sided prism when you look at the crystal on one end, and the cross section is hexagon.

1.3.1.1.5 Trigonal lattice structure: Trigonal structure has a single threefold axis of rotation instead of the sixfold axis of the hexagonal division.

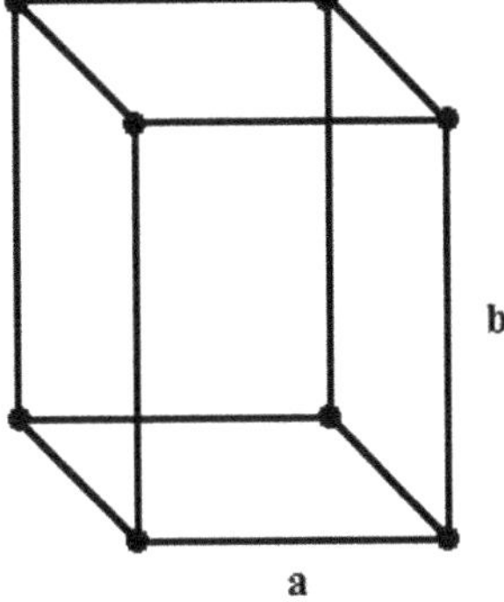

Fig. 1.2 Tetragonal structure

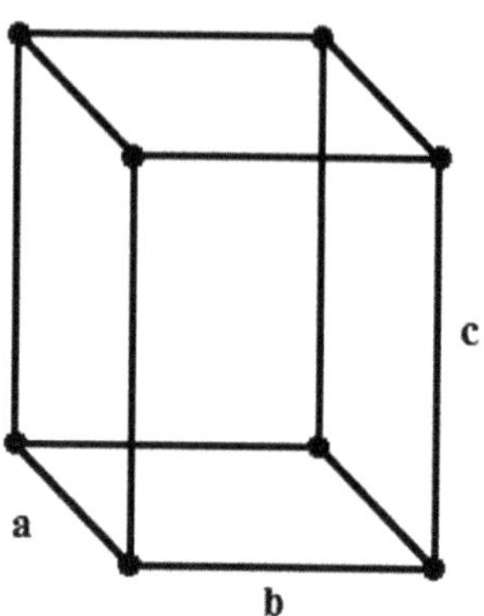

Fig. 1.3 Orthorhombic structure

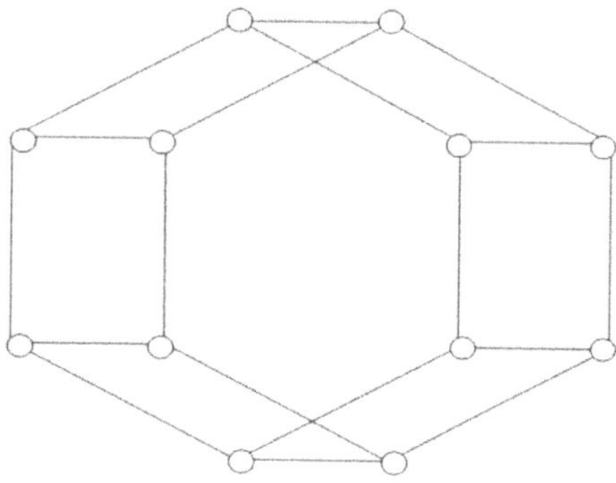

Fig. 1.4 Hexagonal cubic structure

1.3.1.1.6 Triclinic lattice structure: Triclinic structure is usually not symmetrical from one side to the other, which can lead to some fairly strange shapes.

1.3.1.1.7 Monoclinic lattice structure: These are like skewed tetragonal (structure) crystal of a forming prism and double pyramids.

1.3.2 Types of Solids

There are basically three types of solids such as crystalline, amorphous, and polycrystalline.

1.3.2.1 Crystalline: A crystalline form of solid has periodically repeated arrangements of atoms as shown in Fig. 1.5.

1.3.2.2 Amorphous: This kind of solid has non-period structure, means it contains non-periodic structure as shown in Fig. 1.6.

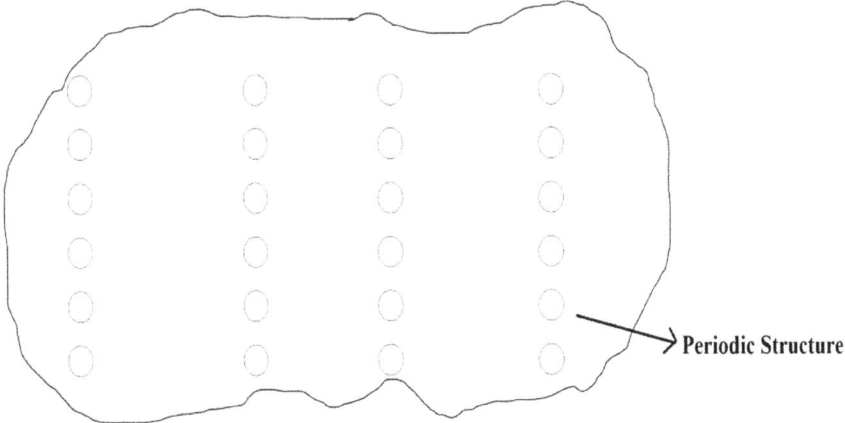

Fig. 1.5 Crystalline structure

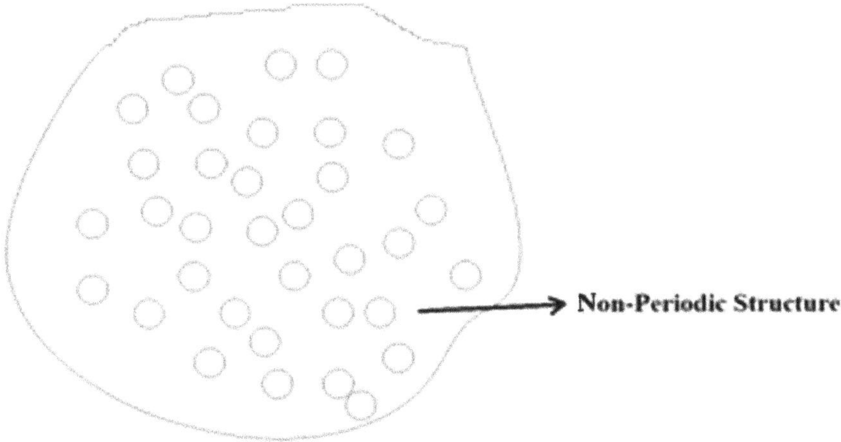

Fig. 1.6 Amorphous structure

1.3.2.3 Polycrystalline: The polycrystalline solids are composed of many small regions of single crystal material as shown in Fig. 1.7.

1.4 Origin of Magnetic Dipoles in Solids

The magnetic dipole shows a metal having two poles named as south and north poles, and it carries the property of magnet called magnetic dipole. The magnetic dipoles coming from a magnetic dipole moment and which is generated by flow of current in a closed circuit/loop now, the magnetic moment.

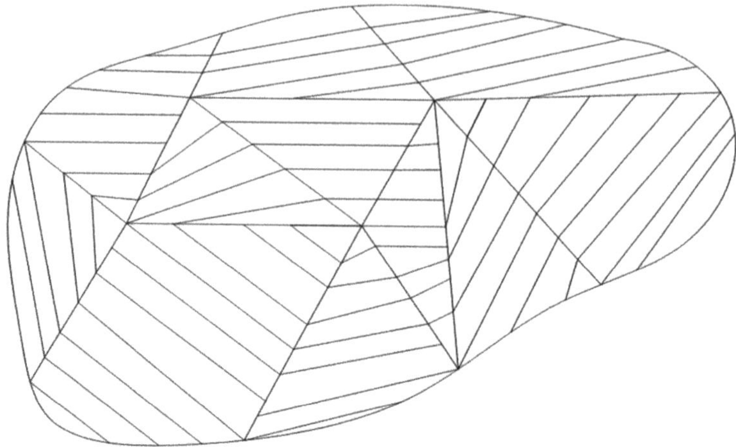

Fig. 1.7 Polycrystalline structure

$$m = I \times A \tag{1.1}$$

where *I* represents the flow of current. *A* is area of closed circuit. According to Bohr's model for an atom, if an electron rotation around the nucleus at a distance "*r*" having angular frequency "*v*" and carried a current "*I*," then the value of carried current "*I*" will be

$$I = e \times v \tag{1.2}$$

where *v* is angular frequency, where $v = \frac{w}{2\pi}$

Now the value of

$$I = e \times \frac{w}{2\pi} \tag{1.3}$$

Now the area of circle is πr^2 so we can say that the magnetic moment "m_{orb}" of the electrons is

$$m_{\text{orb}} = e \times \frac{w}{2\pi} \times \pi \times r^2 \tag{1.4}$$

$$m_{\text{orb}} = \frac{1}{2} \times e \times w \times r^2 \tag{1.5}$$

where *r* is the distance between electron and nucleus. Now, the angular momentum is the quantity of rotation of body, which is the product of mass, angular frequency, and square of the distance of the particle from nucleus.

Angular momentum (*L*)

$$L = m_e \times w \times r^2 \tag{1.6}$$

Using Eqs. (1.5) and (1.6), we can say that

$$w \times r^2 = \frac{L}{m_e} \tag{1.7}$$

$$m_{orb} = \frac{1}{2} \times e \times \frac{L}{m_e} \tag{1.8}$$

In Eq. (1.8), m_{orb} is magnetic dipole moment, L is angular moment / Angular momentum, e is electron particle, and m_e is mass of electron. As we know that the direction of magnetic moment and angular moment is anti-parallel. Then,

$$m_{orb} = -\frac{1}{2} \times e \times \frac{L}{m_e} \tag{1.9}$$

Minus (−) sign because mechanical angular momentum "L" and magnetic moment "m_{orb}" are anti-parallel, m_{orb} and L both are vector".

1.5 Magnetic Material

Materials may be classified by this response to external applied magnetic fields that is called magnetization of a material.

Magnetization of a material can be expressed in terms of density of net magnetic dipole moment "m" in a material. We define a vector quantity called the magnetization "M" by $m = \frac{\mu_{total}}{v}$ then the total magnetic field "B" in the material is given by $B = (B_0 + \mu M)$. Where B_0 is externally applied magnetic field and μ_0 is magnetic permeability of free space. When magnetic field inside of materials is calculated using ampere law, then μ_0 is replaced by μ, where $\mu = K_m$, where μ is the relative permeability. The material does not respond to the external magnetic field by producing any magnetization then $K_m = 1$.

Another commonly used magnetic quantity is magnetic susceptibility which specifies how much the relative permeability differs from one magnetic susceptibility (X_m), $X_m = K_m - 1$.

Magnetic moment Magnetic moment of a magnet is a quantitative that determines the force that the magnet can exert on electric current and the torque that a magnetic field exerts on it.

Diamagnetic, paramagnetic, ferromagnetic, etc., are basically the magnetic responses that differ greatly in strength. Diamagnetism is a property of all material and opposes applied magnetic field, but it is very weak.

1.5.1 Types of Magnetization/Magnetism

There are basically five types of magnetism given as

1.5.1.1 Paramagnetic.
1.5.1.2 Diamagnetism.
1.5.1.3 Ferromagnetism.
1.5.1.4 Anti-ferromagnetism.
1.5.1.5 Ferrimagnetism.

1.5.1.1 Paramagnetic

In paramagnetic, in the absence of an external magnetic field, the orientations of magnetic dipoles are random as shown in Fig. 1.8. While applying an external magnetic field, the direction of dipoles is in the same direction or in the parallel with external magnetic field.

For some solid materials, each atom possesses a permanent dipole moment by virtue of incomplete cancellation of e^- spin and/or orbital magnetic moments. In the absence of an external magnetic field, the orientations of these atomic magnetic moments are random, such that a piece of material possesses no net macroscopic magnetization; these atomic dipoles are "free to rotate," and paramagnetism results when they preferentially align, by rotation, with an external field as shown in Fig. 1.9.

These magnetic dipoles are acted on individually with no mutual interaction between adjacent dipoles. In as much as the dipoles align support with the external field, they enhance it, and $\mu_r > 1$.

Example: Al, Cr (chromium), Mn (manganese), etc.

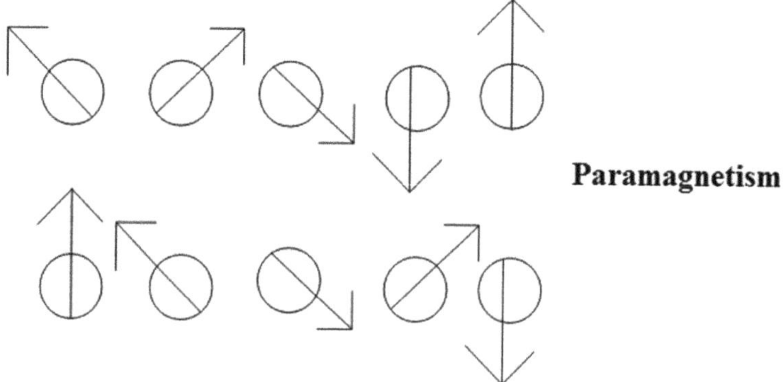

Fig. 1.8 Paramagnetic structure

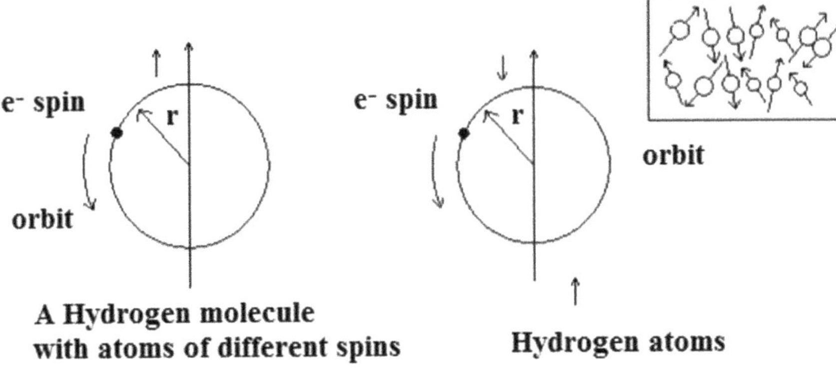

Fig. 1.9 A hydrogen molecule with atoms of different spins

For paramagnetic and diamagnetic materials, the relative permeability is very close to 1, and the magnetic susceptibility is very close to 0. For ferromagnetic material, these quantities may be very large.

- Some materials like Al and Pt exhibit a magnetization which is proportional to the applied magnetic field in which the material is placed.
- These materials are set to be paramagnetic.
- All atoms have inherent source of magnetism because electron spin contributes a magnetic moment and electron orbit acts as current loops which produce a magnetic field.
- In most materials, the magnetic moment of the electrons is cancelled, but in materials which are classified as paramagnetic, the cancellation is incomplete.

1.5.1.2 Diamagnetism

Diamagnetism is a very weak form of magnetism. In this magnetism, we do not apply any external field or absence of an external field. No dipole exits. While an external field is being applied, then the dipoles are introduced and showed dipole direction is opposite to the external magnetic field direction as shown in Fig. 1.10a. This change in direction between dipole and external field introduces by a change in the orbital motion of electrons due to applied magnetic field. That is why dipole direction is opposite to that of the applied field.

- The orbital motion of electrons creates tiny atomic current loops which produce magnetic fields as shown in Fig. 1.11.
- When an external magnetic field is applied to a material, these current loops will tend to align such a way as to oppose the applied field.
- This may be viewed as an atomic version of Lenz's law (induced magnetic field tends to oppose the change which created them).

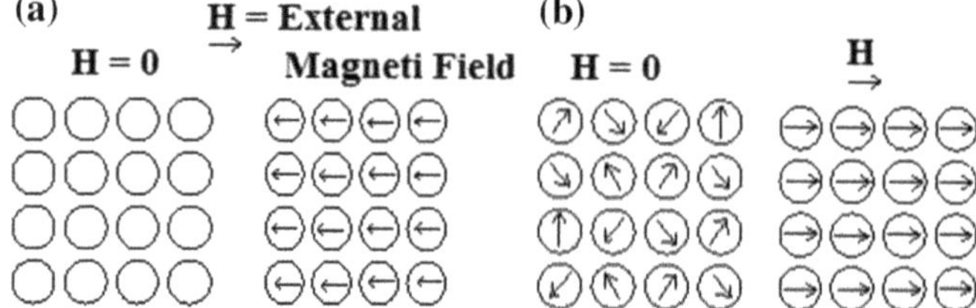

Fig. 1.10 **a** Atomic dipole configuration for a diamagnetic material with and without a magnetic field. In the absence of an external field, no dipole exists; in the presence of a field, dipoles are induced that are aligned opposite to the field direction. **b** Atomic dipole configuration with and without an external magnetic field for a paramagnetic material

Fig. 1.11 Forces acting on an e^-

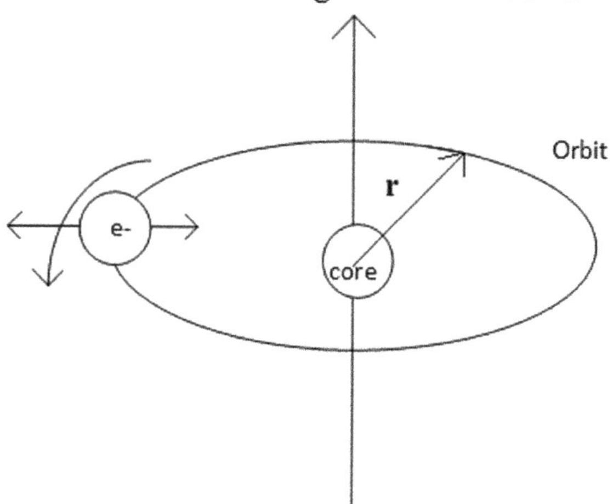

- Materials in which this effect is the only magnetic responses are called diamagnetic materials.
- All materials are inherently diamagnetic, but if the atoms have some net magnetic moment as in paramagnetic material or if there is a long range ordering of atomic magnetic moments as in ferromagnetic material, then these stronger effects are always dominant.
- Diamagnetism is the residual magnetic behavior; then materials are like a paramagnetic and non-ferromagnetic.
- Any conductor which shows strong effect is called diamagnetic effect in the presence of changing magnetic field because circulating current will be generated in the conductor to oppose the magnetic field changes. A superconductor will be a perfect diamagnetic since there is no resistance to the forming of the current.

1.5.1.3 Ferromagnetism

Figure 1.12 depicts the structure of ferromagnetism. In ferromagnetism, it holds a permanent magnetic moment in the absence of external magnetic field as shown in Fig. 1.13. The atomic moments in these materials show very strong interactions. These interactions are produced by electronic exchange forces and result in a parallel alignment of atomic moment. While applying an external magnetic field, the directions of dipoles are in the same directions of a magnetic field strongly.

Certain metallic materials possess/hold a permanent magnetic moment in the absence of an external field, and manifest very large and permanent magnetizations. These are the characteristics of ferromagnetism, and they are displayed by the transition metal iron, cobalt, Ni.

The atomic moments in these materials exhibit very strong interaction. These interactions are produced by electronic exchange forces and result in parallel or antiparallel alignment of atomic moments, and exchange forces are very large, equivalent to a field on the order of 1000 T, or approximately a 100 million times the strength of the earth's field.

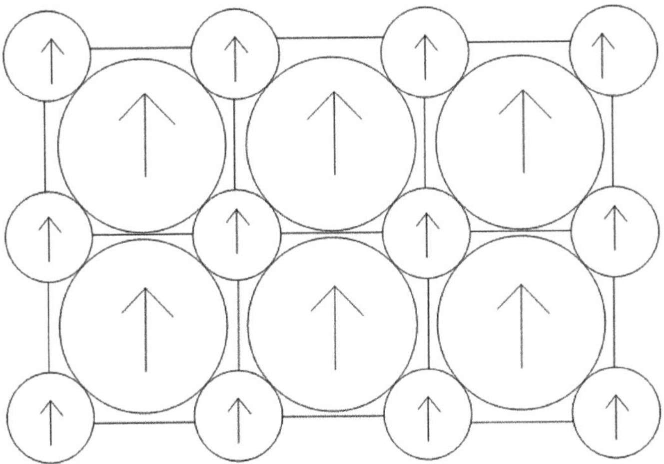

Fig. 1.12 Ferromagnetism structure

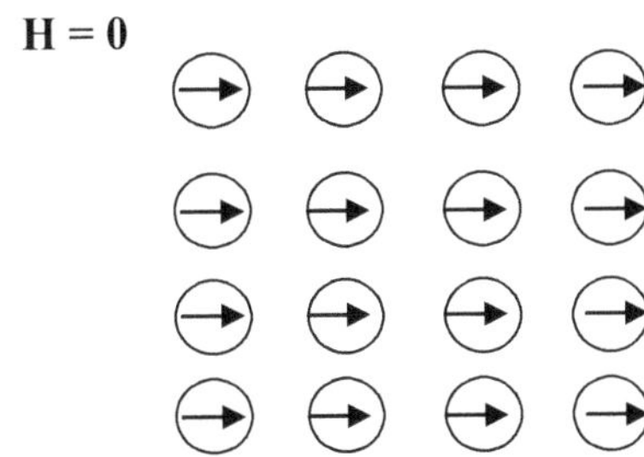

Fig. 1.13 Schematic illustration of the mutual alignment of atomic dipoles for a ferromagnetic material which will exist even in the absence of an external magnetic field

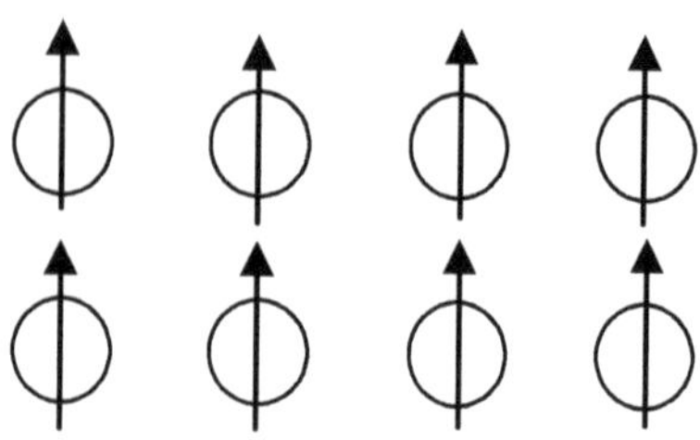

Fig. 1.14 Schematic illustration of the mutual alignment of atomic dipoles for a ferromagnetic material

Ferromagnetic materials exhibit parallel alignment of moments resulting in large net magnetization even in the absence of a magnetic field as shown in Fig. 1.14.

Figure 1.14 shows the mutual alignment of atomic dipoles for a ferromagnetic material, which will exist even in the absence of an external magnetic field.

1.5.1.4 Anti-ferromagnetism

Figure 1.15 depicts the structure of anti-ferromagnetism. In this anti-ferromagnetic, the magnetic moment coupled between adjacent atoms or ions occurs in metals. The coupling results are an anti-parallel alignment; the alignment of the spin moment of atoms or ions is exactly in opposite direction because sublattices *A* and *B* are in opposite direction having same equal magnitude. Thus, the net magnetic moment is 0 (zero).

This type of magnetic ordering is known as anti-ferromagnetism. For example, manganese oxide (MnO) is one material that displays this behavior. MnO is a ceramic material that is ionic in character having both Mn^{2+} and O^{2-} ions.

If the *A* and *B* sublattice moments are exactly equal but opposite, the net moment is zero "0"; this type of magnetic ordering is called "anti-ferromagnetism." This phenomenon of magnetic moment coupling between adjacent atoms or ions occurs

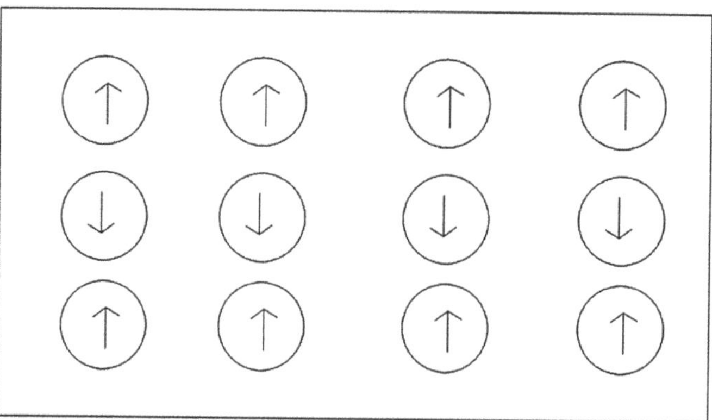

Fig. 1.15 Anti-ferromagnetism structure

in materials other than those that are ferromagnetic. In one such group, this coupling results in an anti-parallel alignment, and the alignment of the spin moments of neighboring atoms or ions in exactly opposite directions is termed "anti-ferromagnetism".

Example: Manganese oxide (MnO).

1.5.1.5 Ferrimagnetism

In ferrimagnetism, it is an ionic compound such as oxides, more complex form of magnetic ordering can occur as a result of the crystal structure. The magnetic structure is composed of two magnetic sublattices (called *A* and *B*) separated by oxygen (O_2). In ferrimagnetism, the magnetic moment of *A* and *B* sublattices is not equal, i.e., opposite direction of each other with a different amount as shown in Fig. 1.16.

In ionic compounds, such as oxides, more complex forms of magnetic ordering can occur as a result of the crystal structure. One type of magnetic ordering is called ferrimagnetism, and a simple representation of the magnetic spins in a ferromagnetic oxide is shown here.

The magnetic structure is composed of two magnetic sublattices (called *A* and *B*) separated by oxygen "(O_2)."

The exchange interactions mediated the oxygen anions; when this happens, the interactions are called indirect or superexchange interactions.

In ferrimagnetism, the magnetic moments of the *A* and *B* sublattices are not equal and result in a net magnetic moment. Ferrimagnetism is therefore similar to ferromagnetism.

Example: Fe_3O_4.

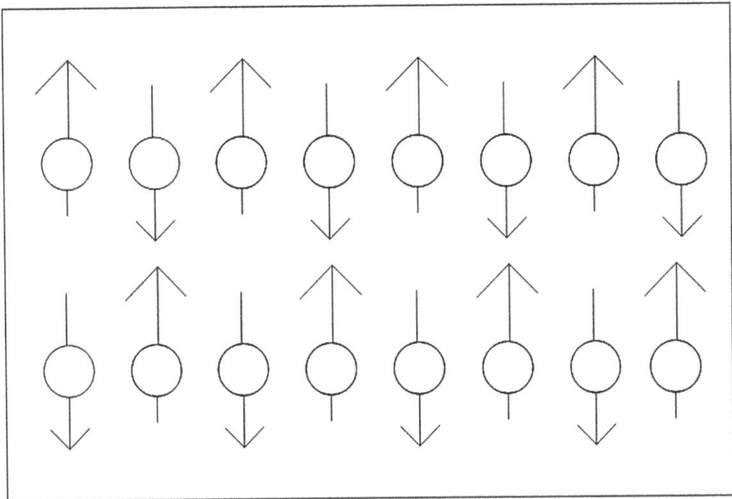

Fig. 1.16 Ferrimagnetism structure

Summary

- **This chapter included introduction of semiconductor materials and discussed various band gaps.**
- **Classify semiconductor materials and discuss them.**
- **Different types of compound materials.**
- **Discuss crystal lattice structure and their classifications.**
- **Derive mathematical expression for origin of magnetic dipoles in solids.**
- **Magnetic materials and magnetization.**
- **Discuss various magnetizations like paramagnetic, diamagnetism, ferromagnetism, anti-ferromagnetism, and ferrimagnetism.**

Numerical Problems

Question 1: Describe semiconductor material and its classifications.
Question 2: State and prove the derivation of origin of magnetic dipoles in solids.
Question 3: Explain magnetic properties of materials.
Question 4: What is magnetization? Write down about paramagnetism and ferromagnetism.
Question 5: What are the types of solids? Explain.

Chapter 2
Special Diodes and Linear Wave Shaping

Learning Objectives

- **Discuss various diodes like LED, varactor diode, photodiode, Schottky diode, and tunnel diode with their working operation.**
- **Discuss advantages and disadvantages of various diodes with their applications.**
- **Introduce various filters, LPF, PF, BPF, and BSF.**
- **Define integrator and differentiator.**
- **Explain compensated attenuator.**

2.1 Introduction

Number of diodes having single P–N junction but having different modes of operation, different methods of construction, different characteristics, and special areas of application are available. Such diodes are called special diodes. Some of such special diodes are zener diode, varactor diode, Schottky diode, tunnel diode, photodiode, light emitting diode, and so on.

In this chapter, we discuss various kinds of diodes, their working operations, merits, demerits, and applications.

G. S. Tomar and A. Bagwari, *Fundamentals of Electronic Devices and Circuits*, Algorithms for Intelligent Systems, https://doi.org/10.1007/978-981-15-0267-5_2

2.2 Light Emitting Diode (LED)

The light emitting diode (LED) is an optical diode which emits light when forward biased and introduced in 1962. The symbol of LED which is similar to P–N junction diode apart from the two-arrows indicating that the device emits the lights energy as shown in Fig. 2.1.

2.2.1 *Basic Operation of LED*

Whenever the P–N junction is forward biased, the electrons (e^-) across the P–N junction form the N-type semiconductor material and recombine with the holes in the P-type semiconductor material. In Fig. 2.2, the free electrons are in the conduction band while the holes are present in the valance band. Thus, the free electrons are at higher energy level with respect to the holes. When the free electrons with holes fall from conduction band to valance band. Therefore, the energy level associated with it changes from higher value to lower value. The energy corresponding to the difference

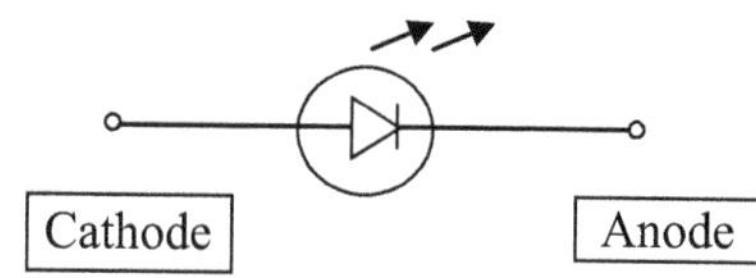

Fig. 2.1 Illustrates symbol of LED

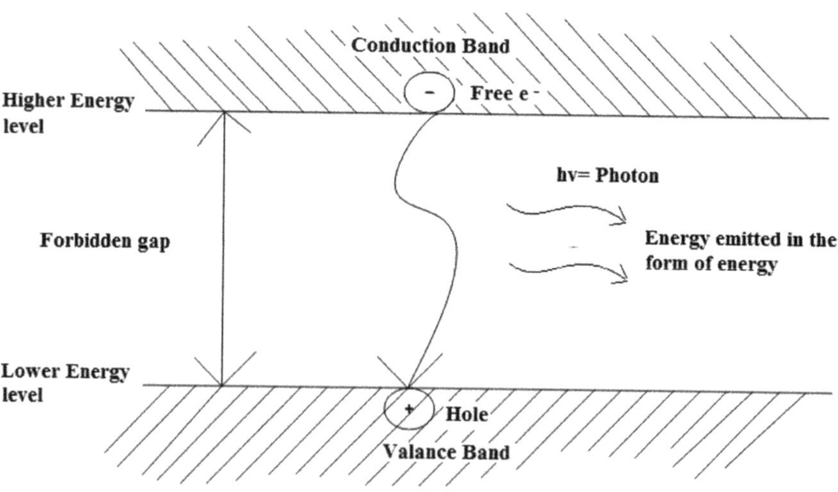

Fig. 2.2 Process of electroluminescence

between higher level and lower level is released by an electron while traveling from the conduction band to the valance band.

In LED, the energy is released in the form of photons which emit the light energy. Hence, such diodes are called light emitting diode (LED) and process is called electroluminescence. Between P-type and N-type, there exists an "active region." This active region emits lights when an e^- and hole recombine during the diode is forward biased. Holes form P-type and e^- form N-type, both get driven into the active region and when recombine, the light is emitted.

Where $\lambda = \frac{c}{\upsilon}$, and $h = 6.67 \times 10^{-34}$ Js.

2.2.2 *Material Used by LED*

LEDs use the material like gallium arsenide (GaAs), or gallium arsenide phosphide (GaAsP), or gallium phosphide (GaP).

1. GaAs—No color or colorless (invisible color).
2. GaAs—Red or yellow (visible color).
3. Gap—Red or green.

2.2.3 *LED Voltage Versus Current Graph*

In the given circuit, resistor (R_S) is the current-limiting resistor. Due to this resistor, the current through the circuit is limited and prevents from exceeding the maximum current rating of the diode.

2.2.3.1 Forward Biased

In Fig. 2.3, the voltage (V) drop across conducting LED is about 2–3 V which is considerably greater than that across a normal silicon (Si) or (Ge) diode. The current range of commercially available LEDs is 10–80 mA. Unless and otherwise specified while analysis the LED circuits, the drop across LED is considered as

$$V_D = 2\text{V, for P} - \text{N junction diode it is } 0.7\text{ V}.$$
$$-V_S + R_S I_S + V_D = 0$$

Current through LED used in circuit is

$$I_S = \frac{V_S - V_D}{R_S} \tag{2.1}$$

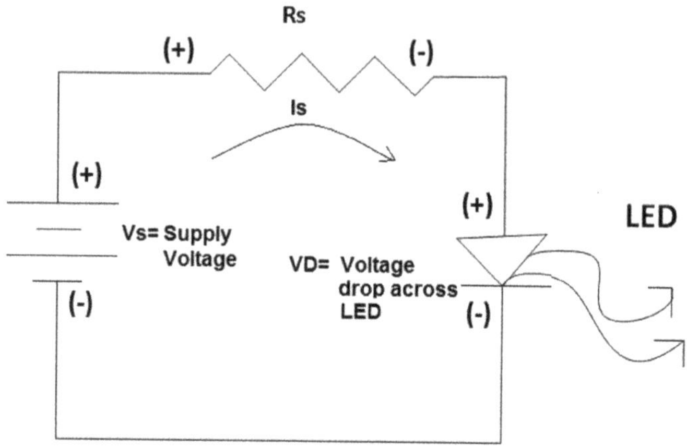

Fig. 2.3 Simple circuitry of LED operation

The reverse breakdown voltage of LED varies 3–10 V. While P–N junction diode >50 V.

2.2.3.2 Output of LED

Figure 2.4 shows the output power versus forward current graph, here the output light proportional to forward current, i.e., more numbers of recombination of P-type and N-type (electrons and holes) recombine generates or emitted more output light.

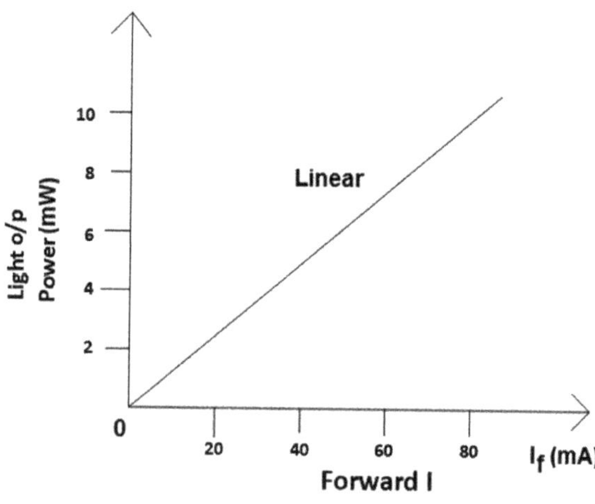

Fig. 2.4 Output power versus forward current graph

2.2.4 *Advantages of LED*

1. Light emitting diode is light in weight
2. LED has long life.
3. LED is cheap and reliable.
4. It is easy interface with varies other electronic circuits.
5. LEDs are available in varies colors.
6. Small size.
7. The brightness of LED depends on the current flow, that is why brightness of LED can be controlled via varying current.
8. It is fast operating devices, on–off or switching time is <1 μs.

2.2.5 *Disadvantages of LED*

1. Temperature affects its characteristics like power, voltage, and current, etc.
2. Need large power for operation compressed to P–N diode.
3. Draws considerable current from battery, i.e., more current draws.

Example 1: Considering Fig. 2.5, compute the current through LED used in circuit.
Answer: Let voltage drop across the LED is 2 V,

Given: $V_S = 15$ V.
$R_S = 22\,\text{k}\Omega$.

$$I_S = \frac{V_S - V_D}{R_S} \Rightarrow \frac{15\,\text{V} - 2\,\text{V}}{2.2\,\text{k}\Omega}$$

$I_S = 5.91\,\text{mA}$.

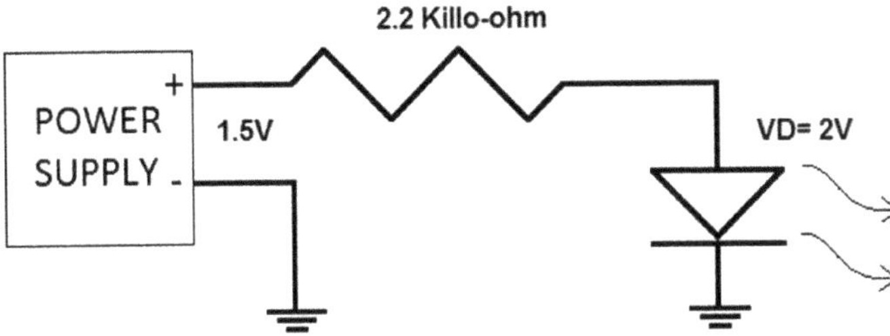

Fig. 2.5 Given circuitry

2.2.6 Application of LED

1. All kinds of visual displays, i.e., seven-segment displays, alphanumeric displays, watch, and calculators.
2. Use in optical devices like OFC for data sending in the form of optical light (LED as a light source).
3. On–off indicator in varies electronics circuits.
4. LEDs are useful in remote controls and applications for visible as well as invisible light transmitted or emission.

2.3 Varactor Diode

Figure 2.6 shows the symbolic representation of the varactor diode. When a diode is reverse biased, the width of the depletion region increases. So there are more positive and negative charges present in the depletion region.

Due to this, the P-region and N-region act like the plates of CAPACITOR "*C*" while the depletion region acts like "di-electric (insulator)." Thus, there exists a capacitor at the P–N junction called transition capacitor, junction capacitor, space charge capacitor, barrier capacitor, or depletion region capacitor denoted by "C_{T}" as shown in Fig. 2.7.

2.3.1 Mathematical Expression

$$C_{\mathrm{T}} = \frac{\mathcal{E} A}{W} \tag{2.2}$$

where $\mathcal{E}$ = Permittivity of semiconductor.
$\mathcal{E}_0 = 8.849 \times 10^{-12}$ Ferrod/m.
$\mathcal{E}_r$ = Relative permittivity of semiconductor (dimensionless).

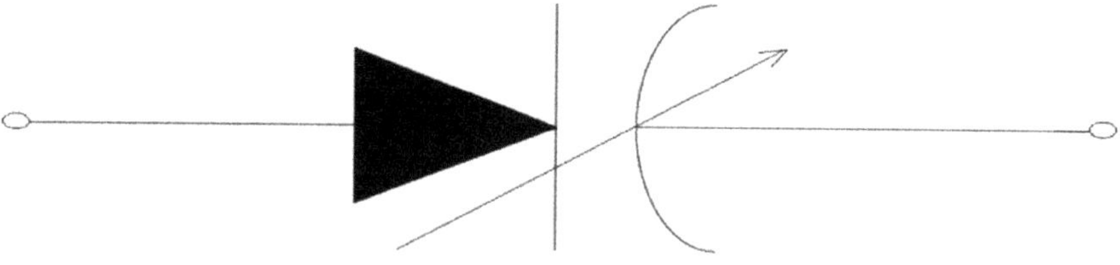

Fig. 2.6 Symbol of varactor diode

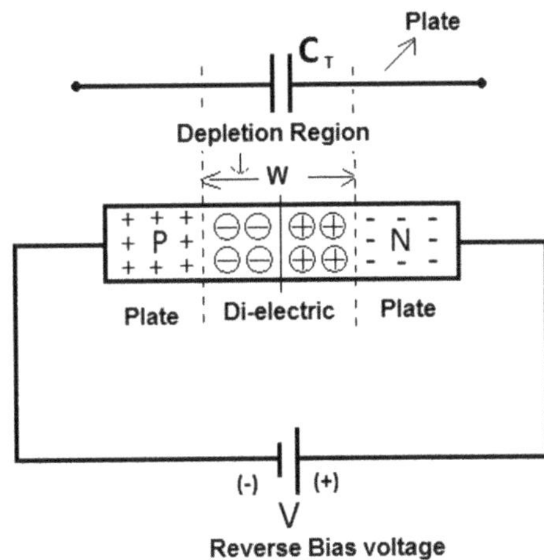

Fig. 2.7 Process of reverse bias forming varactor diode

A = Areas of cross section.
W = Width of depletion region.

After analyzed Eq. (2.2), we can say that depletion region is directly proportional to reverse bias voltage and inversely proportional to transition capacitance.

As the reverse biased voltage increases, the width of the depletion region increases. Thus, the transition capacitance decreases. In short, the transition capacitance (C_{T}) can be controlled by applied voltage.

2.3.2 *Graph Between the Variations of Transition Capacitance (C$_T$) Versus Applied Reverse Bias Voltage (V$_R$)*

In Fig. 2.8, the $C_{\mathrm{T}}(0)$ shows capacitor zero bias condition = 80 pF, and negative or (−)ve shows reverse bias. As the reverse voltage is negative, the graph is shown in second quadrant. For a particular diode shown, C_{T} varies from 80 pF to less than 5 pF as reverse voltage (V_{R}) changes from 2 to 15 V.

As the transition capacitor varies with the applied voltage, it can be used as a "voltage variable capacitance" in many applications. In practice, special types of diodes are manufactured which shows the transition capacitance property more predominantly as compared to the normal diodes. Such diodes are called varactor diodes or varicap or voltage variable capacitance (VVC) or tuning diodes.

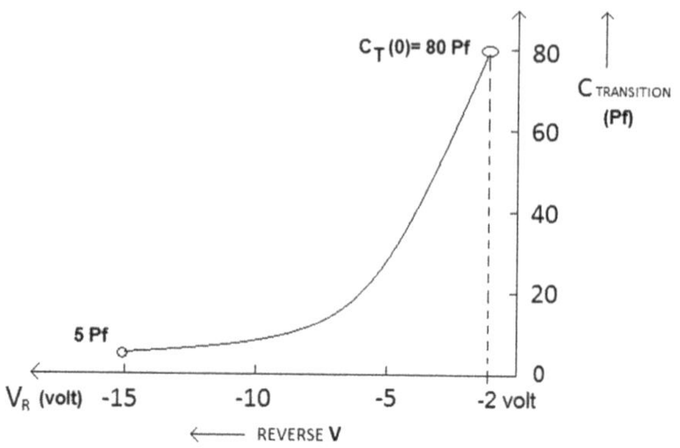

Fig. 2.8 Variation of transition capacitance versus applied reverse bias voltage

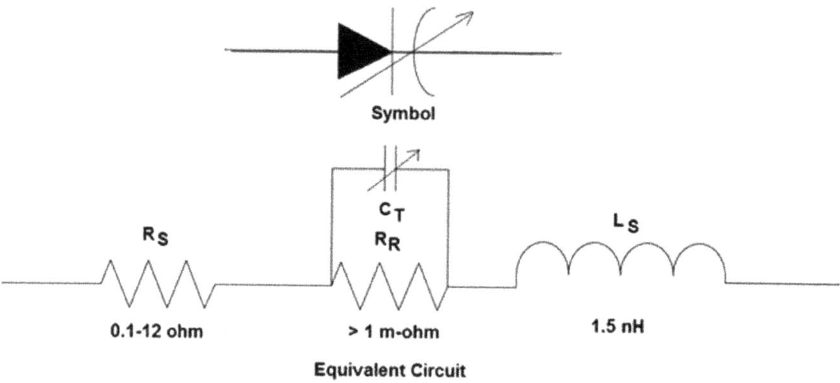

Fig. 2.9 Symbol and equivalent circuit of varactor diode

2.3.3 Symbol and Equivalent Circuit

In the given Fig. 2.9, R_R shows reverse resistor having very large value.

R_S is geometric resistor of diode having very small.
L_S is inductance (High-frequency limit associated).
$f(v)$ represents the functions of voltage.

2.3.4 Expression for Transition Capacitance

For varactor diode, the transition capacitance in terms of applied reverse bias voltage is given by

$$C_{\mathrm{T}} = \frac{K}{(V_{\mathrm{J}} + V_{\mathrm{R}})^n} \tag{2.3}$$

where K shows constant for diode.
V_{J} shows junction potential.
V_{R} shows reverse bias voltage.
n shows diffused junction ($n = 1/2$ for alloy junction, $n = 1/3$ for diffused junction).

Now, under zero biased condition, the capacitance is $C(0)$. Then the transition capacitance can be computed as

$$C_{\mathrm{T}} = \frac{C(0)}{\left(1 + \left|\frac{V_{\mathrm{R}}}{V_{\mathrm{J}}}\right|\right)^n} \tag{2.4}$$

2.3.5 Applications of Varactor Diode

1. Varactor diode used in LC tuned circuit as shown in Fig. 2.10.

$$f_{\mathrm{r}} = \frac{1}{2\pi\sqrt{LC}} \tag{2.5}$$

$$\text{\& standard,}\quad f_{\mathrm{r}} = \frac{1}{2\pi\sqrt{L(C_1 + C_2)}} \tag{2.6}$$

where f_{r} is resonance frequency of parallel LC tuned circuit, which is that frequency where we have maximum amplitude.

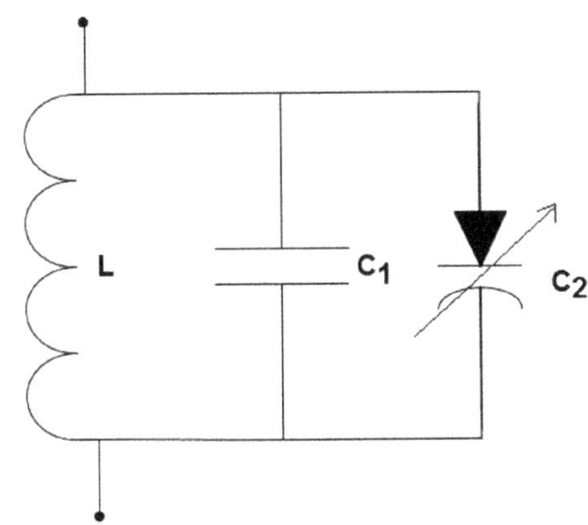

Fig. 2.10 LC tuned circuit

Now, $C_R = C_1 \| C_2$.
and, $C = C_1 + C_2$
$C_2 =$ Transition capacitance of varactor diode.

The C_2 value can be changed by controlling the applied voltage. Hence, circuit can be tuned by changing applied voltage at a resonance frequency, this is called electrical tuning.

2. FM modulation.
3. Television receiver.
4. Automatic frequency control device.
5. Adjustable BPF.
6. Parametric Amplifier.

Example 1: Compute transition capacitance of a diffused junction varactor diode at reverse bias voltage of 4.2 V, if $C(0) = 80$ pF, junction potential of 0.7 V. Also calculate constant "K" for diode.
Answer: (i) The formula for transition capacitance is

$$C_T = \frac{C(0)}{\left[1 + \left|\frac{V_R}{V_J}\right|\right]^n} \quad V_R = 4.2\,\text{V}, \text{ and } V_J = 0.7\,\text{V}.$$

$n = 1/3$ as diffused junction.

$$C_T = \frac{80 \times 10^{-12}}{\left[1 + \left|\frac{4.2}{0.7}\right|\right]^{1/3}}$$
$$C_T = 41.82\,pF.$$

(ii) $C_T = \frac{K}{(V_R + V_J)^n}$

$$41.82 \times 10^{-12} = \frac{K}{(4.2 + 0.7)^{1/3}}$$

and, $K = 71.03 \times 10^{-12}$.

2.4 Photodiode

The photodiode is a semiconductor P–N junction diode whose region of operation is limited to the reverse biased region. Photodiode is also known as a variable resistance diode.

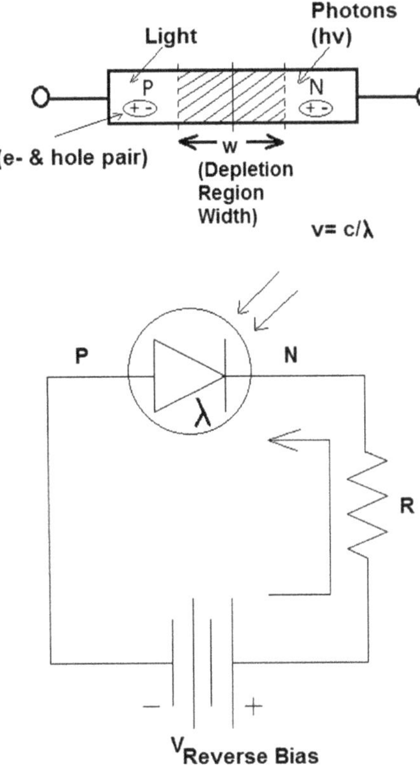

Fig. 2.11 Equivalent circuit of photodiode

2.4.1 Working Principle of Photodiode

In Fig. 2.11, the photodiode is connected in reverse biased condition. The depletion region width is large. Under normal condition, it carries small reverse current due to minority charge carriers. When light is incident through glass window on the P–N junction, photons in the light bombard the P–N junction and some energy is imported to the valence electrons. Due to this, valence electrons are dislodged from the covalent bonds and become for electron. Thus, more electron–hole pairs are generated. Thus, the total number of minority charge carriers increases, and hence, the reverse current increases. This is the basic principle of operation of photodiode.

2.4.2 Characteristics of Photodiode

The photodiode is designed such that it is sensitive to the light. When there is no light, the reverse biased photodiode carries a current which is very small and is called dark current, denoted as (I_λ) in Fig. 2.12a. It is purely due to thermally generated minority

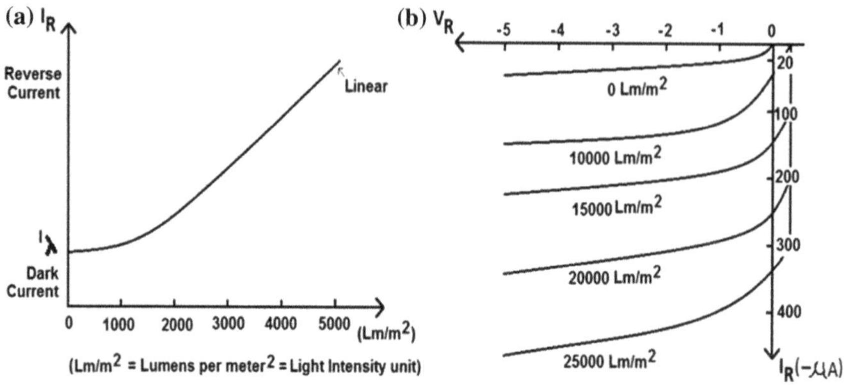

Fig. 2.12 **a** Reverse current versus light intensity curve. **b** Reverse voltage versus reverse current curve

carriers. When light is allowed to full on P–N junction through small window, photons ($h\nu$) transfer energy to valence electron to make them free. Hence, reverse current increases, and it is proportional to light intensity.

Using Fig. 2.12b, we can say that reverse current is "not depended" on reverse voltage and totally depends on light intensity.

2.4.2.1 Use of Photodiode as Variable R-Device

Using graphs, taking $I_\lambda = 20\ \mu\text{A}$ at $V_\text{R} = -2$ V (after that value is changing).

(i) For 0 lumens/m or (Lm/m^2).

$$\text{Dark Resistance} = \frac{V_\text{R}}{\text{Dark} - I(I_\lambda)} \Rightarrow \frac{-2}{20\ \mu\text{A}}$$
$$R \Rightarrow 100\ \text{k}\Omega. \tag{2.7}$$

(ii) For 25,000 Lm/m^2.

Up to $V_\text{R} = -2$ V, $I_\lambda = 350\ \mu\text{A}$.
Then Illuminated $R = \frac{2}{350\ \text{A}}$

$$R \Rightarrow 5.714\ \text{k}\Omega. \tag{2.8}$$

After analyzed Eqs. (2.7) and (2.8), we can say that photodiode can be used as a "variable-*R* device" controlled by light intensity. It is also called "Photo Conductive Device." The response of photodiode is very fast; hence, change in resistance from light to low or otherwise is also very fast. So, it works as a "variable-*R*."

2.4.2.2 Why to Be Used in Reverse Biased (Photodiode)

The reverse current without light in diode is in the range of microampere (μA) like 20 μA to 0 Lm/m^2. The charge in this current due to the light is also in the range of μA. Thus, such a change can be significantly observed in the reverse current. If the photodiode is forward biased, the current flowing through it is in milliampere (mA). The applied forward biased voltage takes the control of the current instead of the light. The change in forward current due to light is negligible and cannot be noticed. The resistance of forward biased diode is not affected by the light. Hence, to have significant effect of light on the current and to operate photodiode as a "R_{variable}" device, it is always connected in reverse biased condition.

2.4.3 *Advantages of Photodiode*

1. Photodiode can be used as variable resistor device.
2. It is highly sensitive to the light.
3. The speed of operation is very high. The switching of current and hence the resistance value from high to low or otherwise is very fast.

2.4.4 *Disadvantages of Photodiode*

1. Dark current (I_λ) is temperature dependent.
2. Overall characteristics are temperature dependent, hence have poor temperature stability.
3. To derive other circuits, we require amplifier due to the low range of current in (μA).

2.4.5 *Application of Photodiode*

1. **Alarm system**: The reverse current "I_R" continues to flow as long as light beam is incident on the photodiode, and when the light is interrupted, the current I_R drops to the dark current level. This initiated the alarm system sending the alarm as shown in Fig. 2.13.
2. **Counting System**: In Fig. 2.14, each item passes the light beam is broken. Thus, reverse current I_R drops to the dark current (I_d) level. This activates the counting mechanism and counter is increased by one.

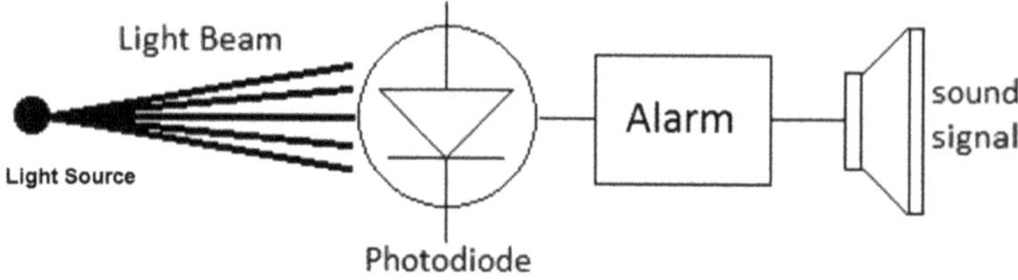

Fig. 2.13 Alarm system

Fig. 2.14 Counting system

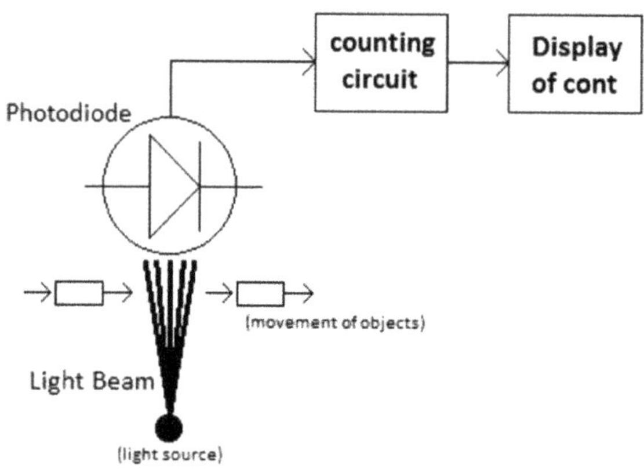

2.5 Schottky Diode

Figure 2.15a shows the symbolic representation of Schottky diode. The Schottky diode, (named after German Physicist) "Walter H. Schottky," is a semiconductor device with a low forward voltage drop and a very fast switching action; when current flows through a diode, there is a small voltage drop across the diode terminals. A normal silicon diode has a voltage drop between (0.6 and 1.7 V). While a Schottky diode voltage drop is between (0.15 and 0.45 V). The lower voltage drop can provide higher switching speed and better system efficiency.

Consider Fig. 2.15b, a metal-semiconductor junction is formed between a metal and a semiconductor, creating a Schottky diode (instead of semiconductor—semiconductor junction as in conventional diodes). Usually, N-type silicon is used as a semiconductor. Different metals, such as molybdenum, platinum, chrome, or tungsten, are used with different construction techniques to get different set of characteristics such as increased frequency range, lower forward bias, etc.

In both the materials metal as well as N-type semiconductor, the electrons (e^-) are the majority carriers. In the metal, the minority carriers (holes) are very less in number. When the contact is made between the two materials, the e^- from the N-type semiconductor material immediately flows into the adjoining metal. This is

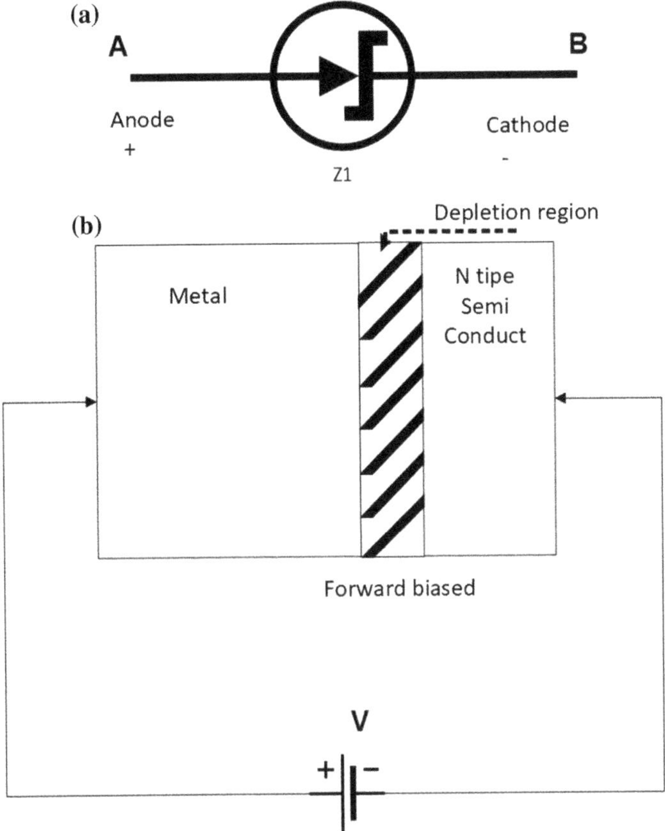

Fig. 2.15 **a** Symbol of Schottky diode. **b** Basic structure of Schottky diode

because the KE (kinetic energy $=1/2\ mv^2$) level of the majority carriers, i.e., e^- in the N-region is higher than the e^- in the metal.

Hence, a heavy flow of majority carriers is established from N-region to the metal. Due to high KE, the injected carriers are called "Hot-carriers."

In conventional diode, the minority carriers get injected into adjoining region while in Schottky diode, majority carriers get injected into metal.

The heavy flow of e^- into the metal creates a region near the junction surface depleted of carriers in the silicon material. This is similar to depletion region in a conventional diode. The additional carriers in the metal establish a negative wall in the metal at the boundary between the two materials. This results in further current. So, there exists a carrier-free region and a negative wall at the surface of the metal.

At the low frequency, the conventional diode can be easily "turn-off" by changing its bias from forward to reverse bias. But at high frequency, conventional diode shows a tendency to store the charge and there is noticeable current in reverse half cycle.

During forward biased, it is not possible for all the carriers in depletion region to recombine. Some carriers exist in depletion region which is not recombined. Now, if the diode is suddenly reverse biased, the carriers existing in depletion region can flow in the reverse direction for some time. But for large lifetime of these carriers, longer is the flows of current in reverse half cycle. Hence, there is a limitation on the frequency range for which a conventional diode can be used.

The time taken by a diode to "turn-off" from its forward biased state is called "Reverse recovery time." For frequency upto 10 MHz, it is very small but above 10 MHz, it is large and put a limit on the use of conventional diode in such high-frequency applications.

The diodes which are specially manufactured to solve this problem of "fast switching" are called "Schottky diodes." Its construction is different than the conventional P–N junction diode. The construction diagram of Schottky diodes is given in Fig. 2.16. These diodes are also called Schottky barrier diodes, surface barrier diodes, or hat carrier diodes.

2.5.1 *Characteristics of Schottky Diode*

Due to the minority carrier-free region, Schottky diode cannot store the charge. Hence, due to lack of charge storage, it can switch off very fast than a conventional diode. It can be easily "switched-off" for the frequencies above 300 MHz. The barrier at the junction for a Schottky diode is less than that of normal P–N junction diode, in both forward and reverse bias region. The barrier potential and break down voltage in forward bias and reverse bias region, respectively, are also less than P–N junction diode.

The barrier potential of Schottky diode is 0.25 V as compared to 0.7 V for normal diode as shown in Fig. 2.17.

2.5.2 *Reverse Recovery Time*

The reverse recovery time is the time when the diode switches from non-conducting to conducting state and vice versa. Where in a P–N diode, the reverse recovery time can be in the order of hundreds of nanoseconds and less than 100 nanoseconds for fast diodes. Schottky diodes do not have a recovery time as there is nothing to recover from (i.e., no charge carrier depletion region at the junction). The switching time is approximately 100 picoseconds (10^{-12} s) for the small signal diodes and up to tens of nanoseconds for special high capacity power diodes. While in P–N junction switching, there is also a reverse recovery current which is high power semiconductor bring increased EMI noise. With Schottky diodes switching essentially instantly with only slight capacitive lording, this is much less of a concern.

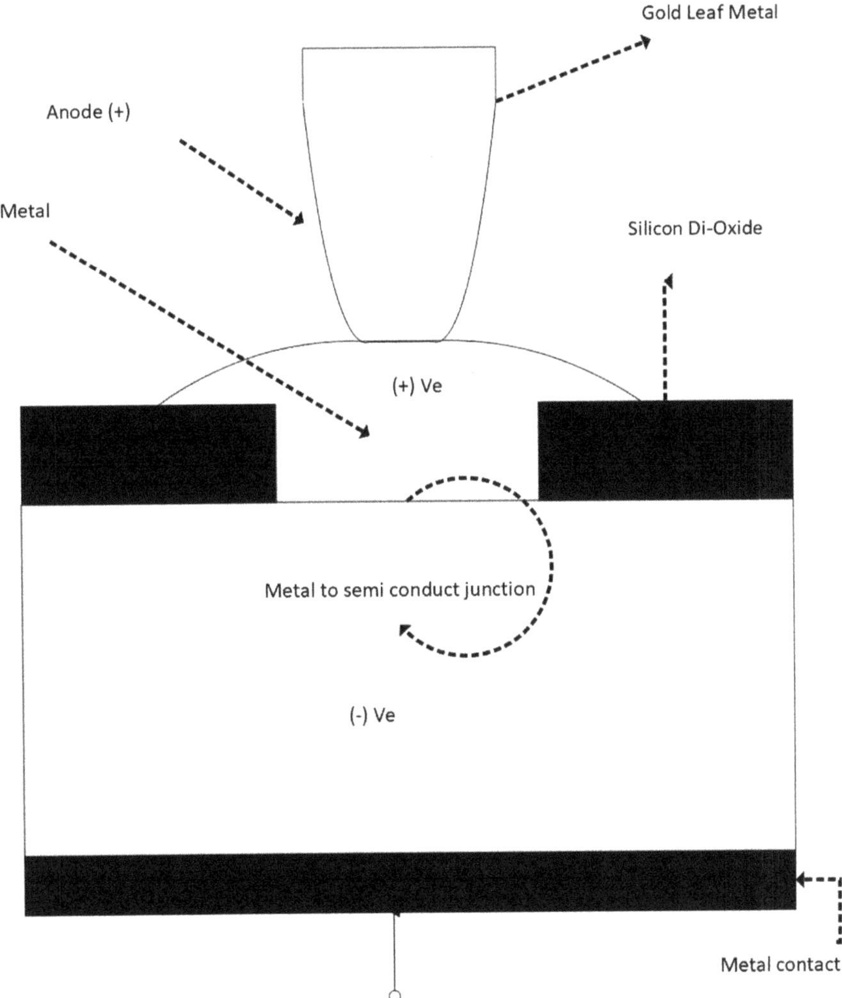

Fig. 2.16 Construction of Schottky diode

2.5.3 *Schottky Diodes Are Useful in Switching*

The high speed of the diode means that the circuit can operate at frequencies in the range 200 kHz–2 MHz. Allowing the use of small inductor (*L*) and capacitor (*C*) with greater efficiency than would be possible with other diode types. Small area Schottky diodes are the heart of "RF detectors and Mixers" which often operate up to 50 GHz.

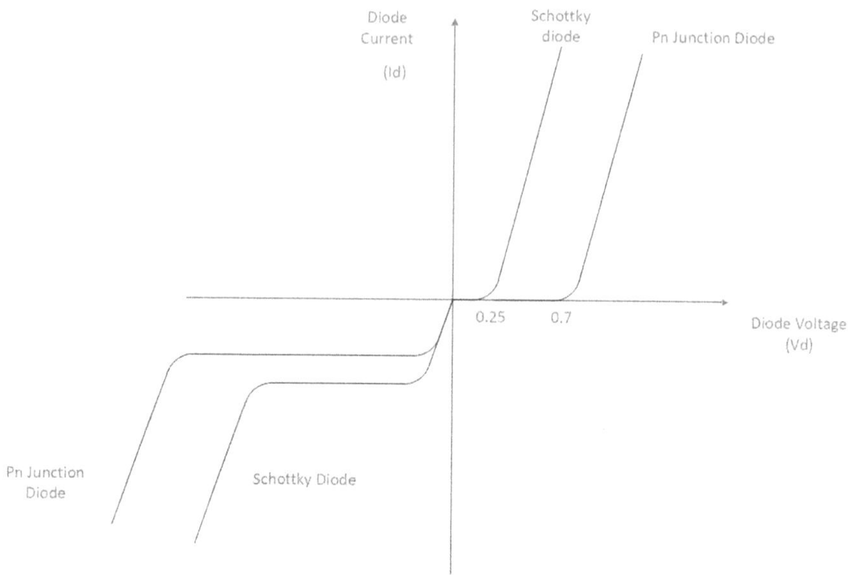

Fig. 2.17 Diode current versus diode voltage graph

2.5.4 *Applications of Schottky Diode*

Due to the fast switching characteristics, the Schottky diodes are very useful for

1. High-frequency applications such as digital computer.
2. High speed TTL.
3. Radar system.
4. Mixers.
5. Detectors in communication equipments.
6. Analog to digital convertor.

2.6 Tunnel Diode

The tunnel diode first developed or reported in 1958 by Japanese scientist "Leo Esaki." Tunnel diode or Esaki diode is a type of semiconductor diode which is capable of very fast operation, well into the microwave frequency region by using "quantum mechanical effects." The symbol of tunnel diode is given in Fig. 2.18.

Quantum mechanical effects, the term is used to describe the manner in which the electrical charges move through the device. Such motion takes place with the speed of light in contrast to the relatively slow motion of electrical charge carriers in transmitter.

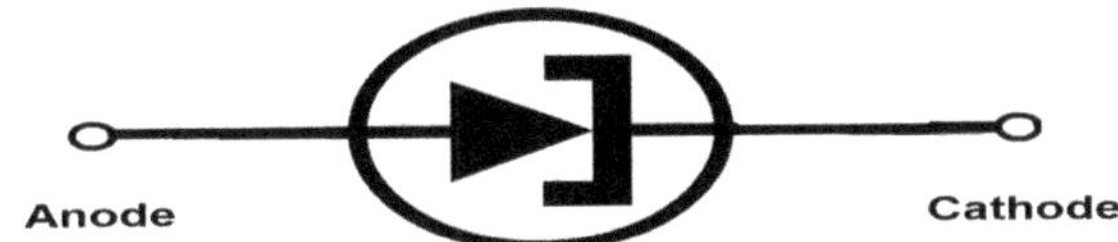

Fig. 2.18 Symbol of tunnel diode

These diodes have a heavily doped P–N junction only some 10 nm (100A) wide. The heavy doping results in a broken band gap, where conduction band e^- states on the N-sides are more or less aligned with valence band hole states on the P-side.

Tunnel diodes are heavily doped and have an extremely narrow depletion or space charge layer of less than 5–10 nanometers thick, only a few atoms deep as shown in Fig. 2.19. In the case of P–N junction diode, the impurity concentration of about 1 part in 10^8 atoms, width $W = 5 \times 10^{-6}$ m. While tunnel diode has impurity concentration of about 1 part in 10^3 atoms, width $W = (5\text{–}10) \times 10^{-9}$ m. Similarly, they are in breakdown when reverse biased, however, unlike the other two devices, they remain in breakdown for a small initial region of forward biasing with the breakdown current superimposed on the normal diode forward current.

Normally, we would not expect breakdown current to flow in the positive region because the electric field is insufficient to overcome the energy gap necessary to release electron into the conduction band; however, the phenomenon is explained by the electron acting as wave rather than as particles.

A certain number of electrons will have more than enough energy to jump an energy gap that would normally be too wide. Effectively, tunneling through a barrier which we would normally expect bar them.

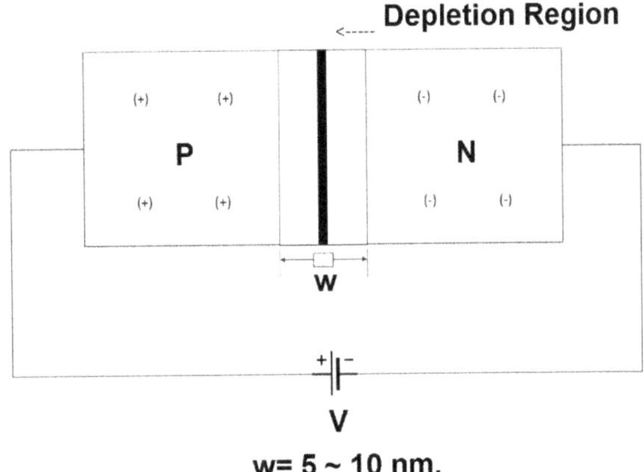

Fig. 2.19 Basic structure of tunnel diode

2.6.1 *Materials Used for Tunnel Diode*

The tunnel diodes are usually made from

1. Germanium (Ge).
2. Germanium arsenide (GgAs).
3. Silicon (Si).

2.6.2 *Operation of Tunnel Diode*

2.6.2.1 Forward Bias Operation

As the voltage beings to increase, electron at first tunnel through the very narrow P–N junction barrier because filled electron states in the conduction band on the N-side become aligned with empty valence band hole states on the P-side of the P–N junction. As voltage increases further, these states become more misaligned and the current drops, this is called "negative resistance" as shown in Fig. 2.20. Because reduction with increasing voltage, as voltage increasing yet further, the diode beings to operate as a normal diode, where electron travels by conduction across the P–N

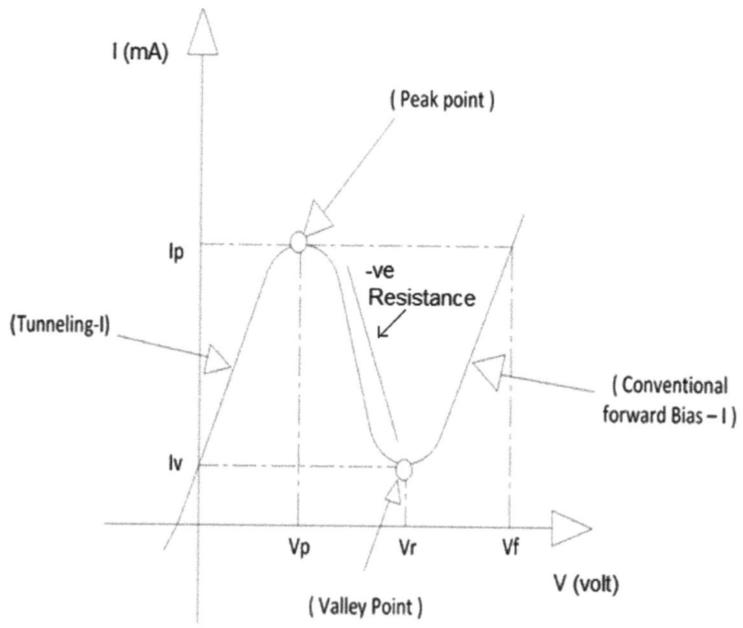

Fig. 2.20 Forward current versus voltage graph

junction, and no larger by tunneling through the P–N junction barrier. Thus, most important operating region for a tunnel diode is "negative resistance region."

2.6.2.2 Reverse Bias Operation

Using in the reverse direction, they are called "Back Diodes"and can act as fast "rectifiers" with zero offset voltage and extreme linearity for power signal (they have on accurate square law characteristics in the reverse direction).

Under reverse biasing, the filled states on the P-side become increasingly aligned, and the empty states on the N-side, the electron move to the tunnel through the P–N junction barrier in the reverse direction, this is the zener effect that also occurs in "Zener Diodes."

2.6.3 *Conventional Diode*

For small forward voltage (up to 50 mV for Ge), the "R" remains small, of the order 5 Ω and current increases. The current attains a peak value (I_P) corresponding to V_P which is about (600 mV). The I_P can vary from few microamperes to several hundred amperes. At the peak point, the slop (dI/dv) of characteristics becomes zero. If now forward voltage is increasing, thus the dynamic conductance (dI/dv) is negative and it shows negative resistance characteristics. This negative resistance continues till a voltage "v_v" called as "volley voltage." At the valley voltage V_v, the I_v and slope $(dI/dv) = 0$. After "V_v," if V_v is increased, the current again increased. Thus, resistance again because positive V_v and remains positive thereafter. After that, we get again "I_p" (Peak Current).

2.6.4 *Construction of Tunnel Diode*

The most common commercially available tunnel diodes are made from the Ge or GeAs. The basic construction of an advanced design tunnel diode is shown in Fig. 2.21.

2.6.5 *Advantages of Tunnel Diode*

1. Tunnel diode has low cost.
2. In tunnel diode, the peak value or point (V_p and I_p) is not a sensitive function of temperature.
3. Tunnel diode has low noise.
4. It has low power consumption.

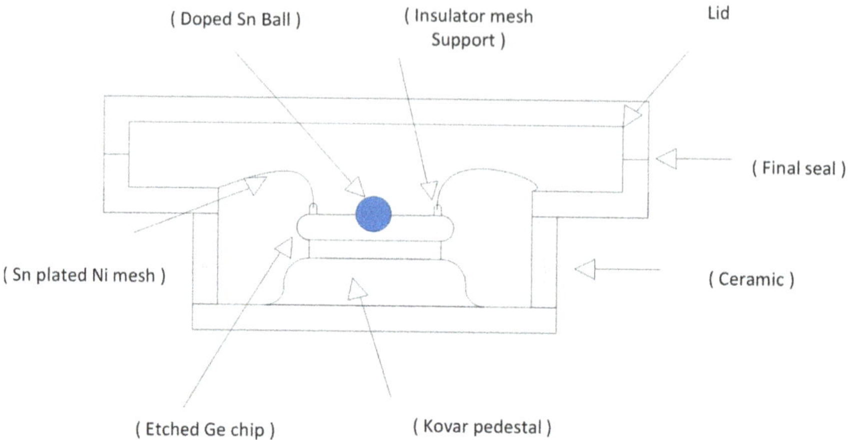

Fig. 2.21 Construction of tunnel diode

5. It has high speed, i.e., the tunneling takes place at the speed of light; hence, the switching times of the order of nanoseconds are easily obtained and switching times as low as 50 picoseconds also can be obtained.

2.6.6 *Applications of Tunnel Diode*

1. High speed computer.
2. Television sets.
3. Communication equipments.
4. Nuclear controls.
5. Satellite and space vehicles.
6. Electronics industries for making electronic equipments (Oscillator, amplifier, frequency converter, and detectors).
7. In pulse and digital circuits.
8. Negative resistance and high-frequency (μ-wave) oscillator
9. Switching networks.
10. In timing and computer logic circuitry.

2.7 Filters

Filter is an electronic device used to remove the unwanted signal from the useful or original signal. It is used to remove or reduce unwanted or noise signal. The oldest

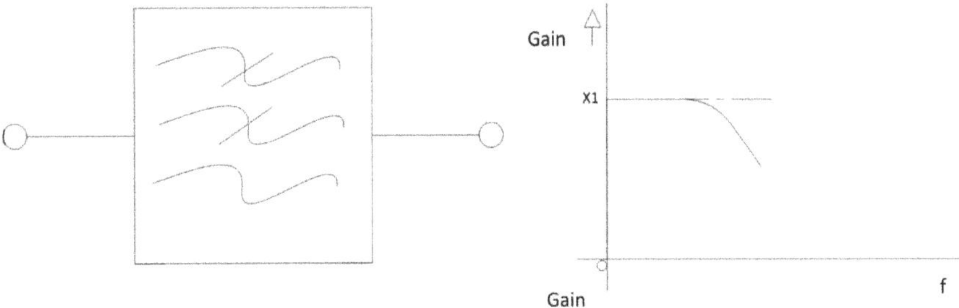

Fig. 2.22 Frequency response of LPF

forms of electronic filters are passive analog linear filter, constructed using resistor (*R*) and capacitor (*C*) or resistor (*R*) and inductor (*L*), known as *RC* or *RL* filter, respectively.

2.7.1 Uses for Filters

Filters are widely used to give circuits such as an amplifier, oscillators, and power supply circuits.

2.7.2 Type of Filters

There are mainly four types of basic filters.

2.7.2.1 Low Pass Filter (LPF).
2.7.2.2 High Pass Filter (HPF).
2.7.2.3 Band Pass Filter (BPF).
2.7.2.4 Band Stop Filter (BSF).

2.7.2.1 Low Pass Filter (LPF)

Low pass filter passes low-frequency signals and blocks other one, i.e., high-frequency signal. Figure 2.22 illustrates frequency response of LPF.

2.7.2.2 High Pass Filter (HPF)

High pass filter passes high-frequency signals and blocks other one, i.e., low-frequency signal. Figure 2.23 illustrates frequency response of HPF.

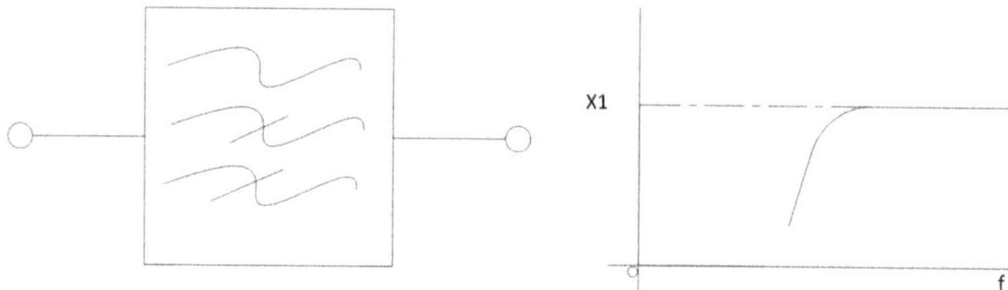

Fig. 2.23 Frequency response of HPF

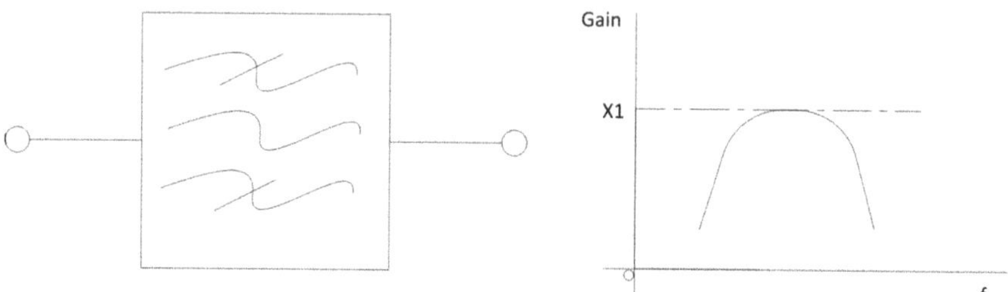

Fig. 2.24 Frequency response of BPF

2.7.2.3 Band Pass Filter (BPF)

Band pass filter passes a particular band of frequencies and blocks others, i.e., beyond the selected band. Figure 2.24 illustrates frequency response of BPF.

2.7.2.4 Band Stop Filter (BSF)

Band pass filter blocks a particular band of frequencies and passes others, i.e., beyond the selected frequency bands. Figure 2.25 illustrates frequency response of BSF.

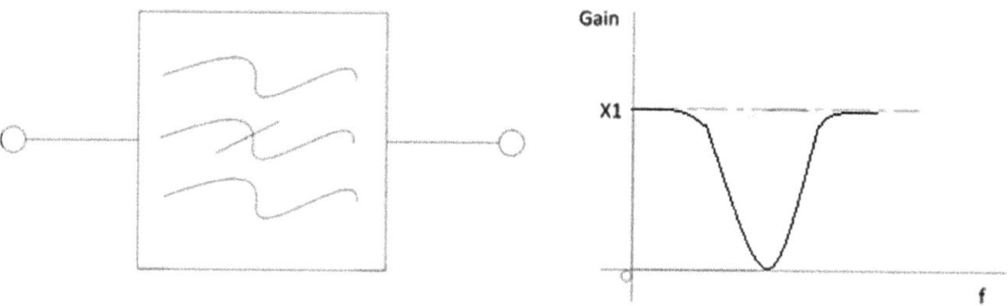

Fig. 2.25 Frequency response of BSF

2.7.3 RC Low Pass Filter and Response to Sine and Square Wave Inputs

2.7.3.1 Low Pass Filter (LPF) with Sine Wave Input

Low pass filter removes or attenuates the higher frequencies signal in circuits such as audio amplifier or we can say it passes the low-frequency signal and stops the high-frequency signals. Figure 2.26 shows the circuit diagram of a low pass filter.

A first under LPF is simply a RC series circuit across the input, with output taken across the capacitor "*C*." We assume that the output of circuit is getting across capacitor *C*.

As we know that at the output $X_c = \frac{1}{2\pi fc}(w = 2\pi f)$
at input $= (R + jX_C)$

$$\begin{aligned}
\text{Then the Gain} = \frac{V_{\text{out}}}{V_{\text{in}}} &= \frac{I * X_c}{I * (R + jX_c)} \\
&= \frac{1/wc}{(R + j * 1/wc)} \\
&= \frac{1/wc}{\sqrt{R^2 + (1/wc)}} \\
&= \frac{1/wc}{\sqrt{\{(Rwc)^2 + (1)\}/wc}}
\end{aligned}$$

The overall gain of LPF is gain $= \frac{1}{\sqrt{(Rwc)^2+1}}$.
The phase of LPF can be written as

$$\begin{aligned}
\theta &= \tan^{-1}\left(\frac{V_R}{V_C}\right) \\
&= \tan^{-1}(RwC).
\end{aligned}$$

Figure 2.27 depicts the waveforms of low pass filter in terms of gain and phase with respect to frequency, respectively.

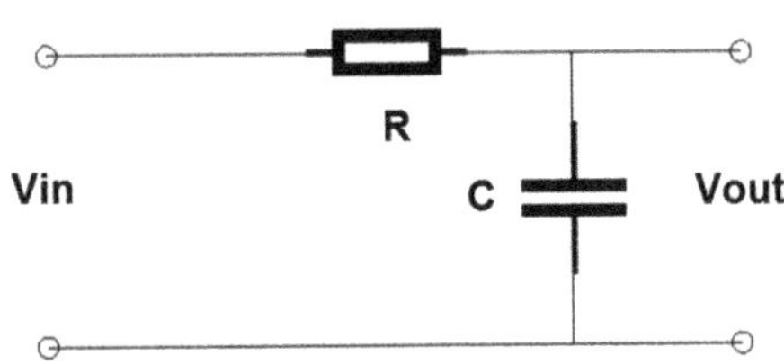

Fig. 2.26 Low pass filter (LPF)

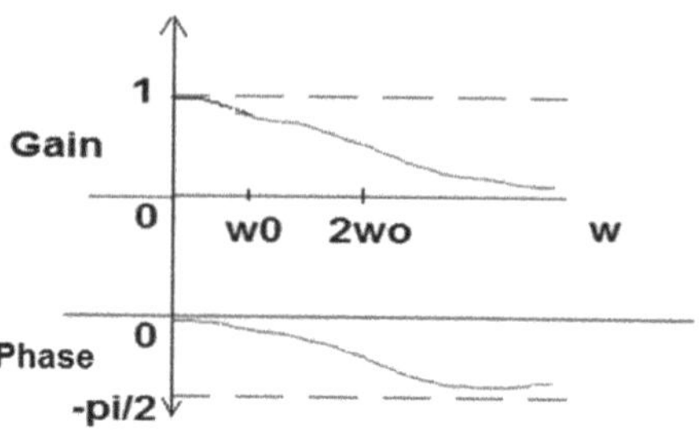

Fig. 2.27 Gain and phase curve of low pass filter

2.7.3.2 Low Pass Filter (LPF) with Square Wave Input/RC Circuit as a Differentiator

When a high pass filter is used with a sine wave input, the output is also a sine wave. The output will be reduced in amplitude and phase shifted when the frequency is low. But it is still a sine wave. This is not the case for square wave input. For non-sinusoidal inputs, the circuit diagram is called differentiator as shown in Fig. 2.28.

Working: Applying square wave as an input to high pass filter (HPF), HPF acts as a differentiator.

Case I: When (RC$\ll T$).

In Fig. 2.29, the output wave is now nothing likes the input wave, but consists of narrow (+)ve spikes and (−)ve spikes. The (+) spikes coincide in time with the rising edge of the input square wave. The (−)ve spikes of the output wave coincide with the falling edge of the input square wave. The circuit called differentiator because its effect is very similar to the mathematical function of differentiation (finding a value that depends on the rate of change of quantity). The differentiator output is effectively a graph of the "rate of change" of the input. Whenever input changing rapidly, a large voltage is produced at the output, and the polarization of output wave depends whether the input is changing in (+)ve and (−)ve spikes coinciding with the rising and falling edges of the input wave.

$$V_{\text{in}} = I * Z \Rightarrow \sqrt{R^2 + (1/wc)^2}, \text{ Let assuming that } (R \ll 1/wc)$$

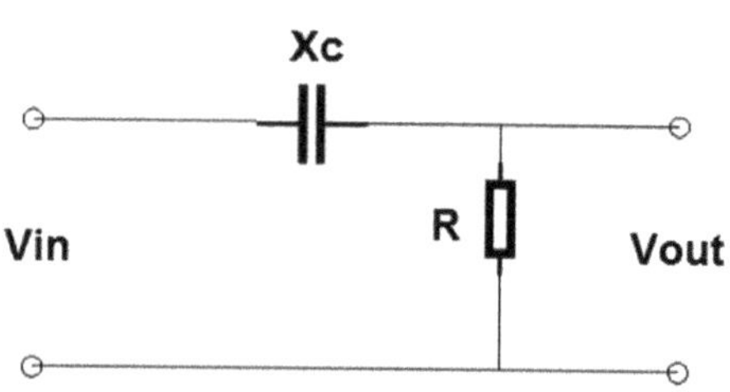

Fig. 2.28 Differentiator having non-sinusoidal signal as an input signal

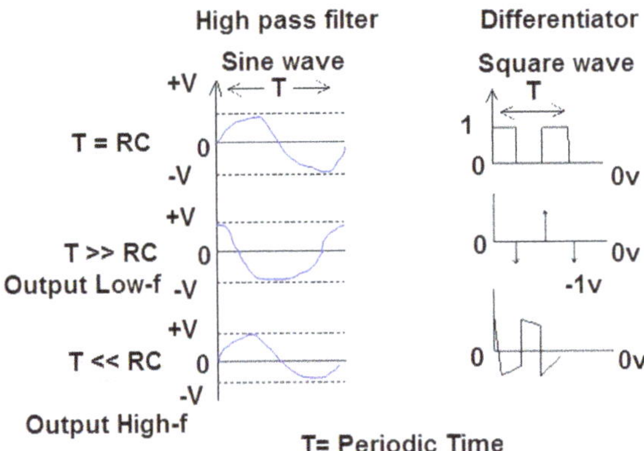

Fig. 2.29 Output waveform of differentiator

So,

$$V_{\text{in}} \cong I/wC$$

for $w \ll \frac{1}{\text{RC}}$, $(T \gg \text{RC})$

$$V_{\text{in}} \cong V_{\text{C}}.$$

$$V_{\text{out}} = V_{\text{R}} = I * R \Rightarrow R\frac{\text{d}}{\text{d}t}q \Rightarrow \frac{\text{d}}{\text{d}t}C * V_{\text{C}}$$

Finally, the output voltage of a differentiator can be calculated as

$$V_{\text{out}} = \text{RC}\frac{\text{d}}{\text{d}t}V_{\text{in}} \tag{2.9}$$

2.7.4 *RC High Pass Filter and Response to Sine and Square Wave Inputs*

2.7.4.1 High Pass Filter (HPF) with Sine Wave Input

In Fig. 2.30, the high pass filter removes or attenuates the lower frequency signals or we can say it passes the high and stop the low-frequency signals. HPF is the combination of C and R. Capacitor "C" is attached at the input side while R at output side. We are taking the output voltage across resistor R.

The overall gain of HPF is gain $= \frac{IR}{I(R+jX_c)} \Rightarrow \frac{RwC}{\sqrt{1+(RwC)^2}}$.

The phase of HPF can be written as $\theta = \tan^{-1}\left(\frac{V_{\text{C}}}{V_{\text{R}}}\right) \Rightarrow \tan^{-1}\left(\frac{1}{RwC}\right)$.

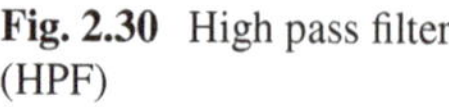

Fig. 2.30 High pass filter (HPF)

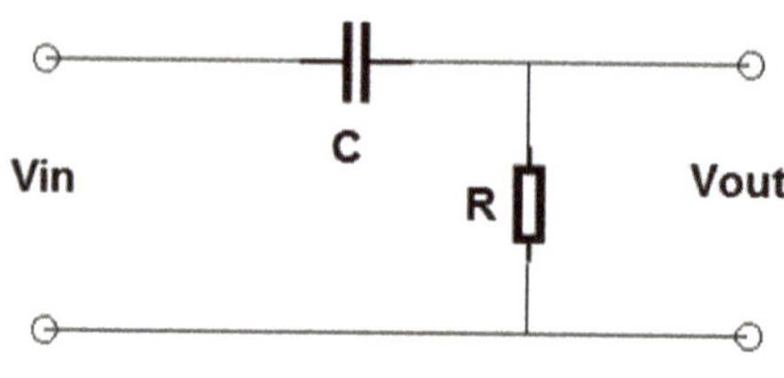

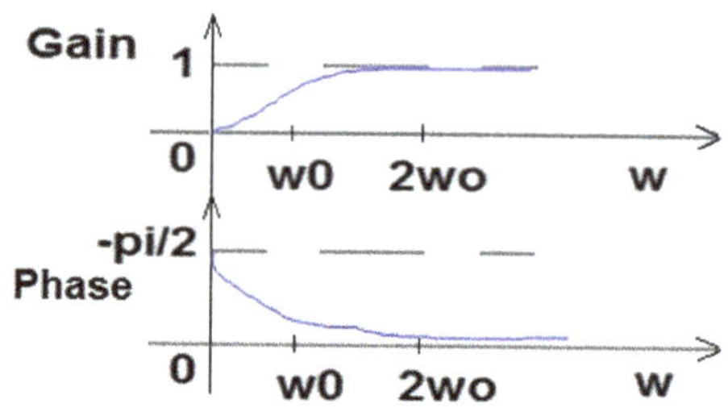

Fig. 2.31 Gain and phase curve of high pass filter

Figure 2.31 depicts the waveforms of high pass filter in terms of gain and phase with respect to frequency, respectively.

2.7.4.2 High Pass Filter (HPF) with Square Wave Input/RC Circuit as an Integrator

In Fig. 2.32, when a LPF is used with a sine wave input, the output is also a sine wave. The output decreases in amplitude and phase shift increases. While for non-sinusoidal input, the circuit is called integrator. The output voltage of an integrator can be calculated as

$$V_{\text{out}} = \frac{1}{\text{RC}} \int V_{\text{in}}\, \text{d}t \tag{2.10}$$

Working: Applying square wave as an input to low pass filter (LPF), LPF acts as an integrator. Integrator has opposite effect to the differentiator. For square wave, output is triangular wave.

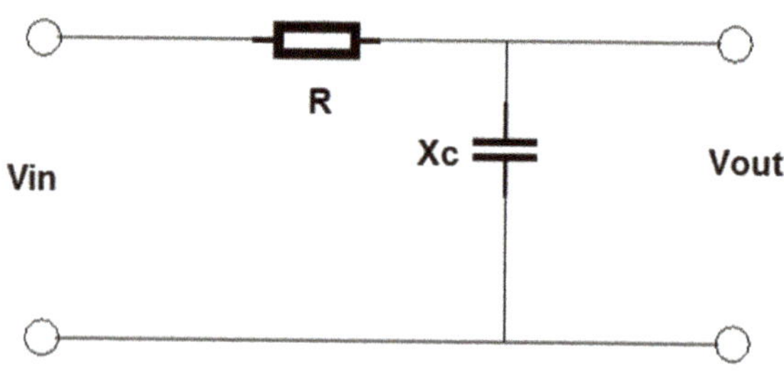

Fig. 2.32 Integrator having non-sinusoidal signal as an input signal

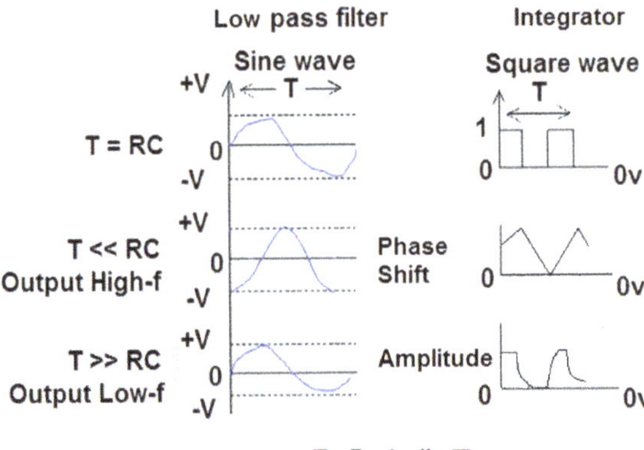

Fig. 2.33 Output waveform of integrator

Case I: When ($T \ll \mathrm{RC}$).

Considered the Low pass filter (LPF), analyzed Fig. 2.33, there are two outputs (1) at low frequency and (2) at high frequency. As we know that LPF passes the low-frequency signal that's why the output of LPF under the low frequency is same or similar as the input signal (only the amplitude changes/affected).

But for high-frequency signal, the output wave is total or not similar to input signal (changing in phase or amplitude).

2.8 Compensated Attenuator

A compensated attenuator is a simple two-port network providing fixed voltage attenuation over a wide frequency band. Such networks are used where precision measurements voltage is required, particularly when the signal is not sinusoidal.

The simplest voltage attenuator is a purely resistive voltage divider with transfer function

$$H(jw) = \frac{V_2}{V_1} \tag{2.11}$$

$$H(jw) = \frac{R_2}{(R_1 + R_2)} \tag{2.12}$$

The transfer function of a resistive voltage divider is independent of frequency only if the resistors exhibit ideal behavior and if parasitic capacitances are negligible. Unfortunately, at high frequencies, a resistor can exhibit substantial parasitic inductance. This series inductance makes the attenuation factor dependent upon the frequency. Another problem evident at higher frequencies is that stray (parasitic) capacitance interferes with the function of a resistive voltage attenuator. A simple way to alleviate this problem is to introduce capacitance in parallel to the resistors.

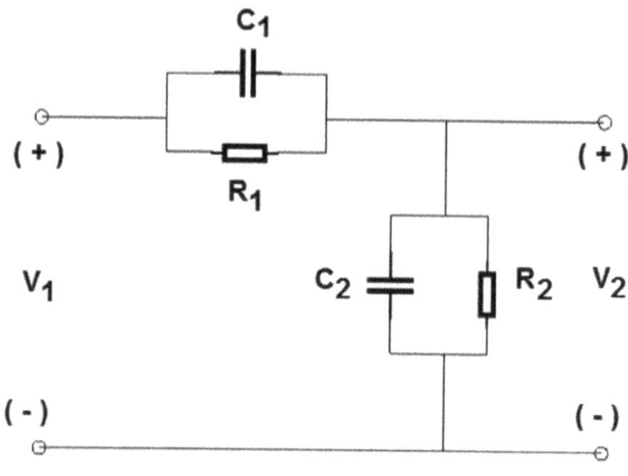

Fig. 2.34 Compensated attenuator

Examine the circuit in Fig. 2.34. It should be obvious that this circuit, known as a compensated attenuator, works like a resistive voltage divider at low frequencies and like a capacitive voltage divider at high frequencies. In essence, the capacitors C_1 and C_2 provide a shunt path for most of the current flow at high frequency, thus overwhelming any series inductance in the resistors as well as any stray capacitance due to the circuit layout, probe configuration, etc.

Summary

- **This chapter included various diodes.**
- **Discuss working operation of LED, features, drawbacks, and applications.**
- **Discuss LED voltage versus current graph.**
- **Working of Varactor diode and applications.**
- **Expression for transition capacitance for varactor diode.**
- **Discuss Photodiode, its characteristics, and applications.**
- **Schottky diode working, its characteristics, and applications.**
- **Tunnel diode working, operations like forward bias, and reverse bias, and their application.**
- **Graph between forward current and voltage for tunnel diode.**
- **Introduce and discuss various types of filters, LPF, HPF, BPF, and BSF.**
- **Integrator and differentiator circuits.**
- **Compensated attenuator.**

Numerical Problems

Question 1: Explain light emitting diode (LED) with its working operation.
Question 2: Write down the difference between LED, varactor diode, and photodiode.
Question 3: What is the explain LED with its working operation.

Question 4: How differentiator is different from high pass filter (HPF).
Question 5: Explain compensated attenuator.
Question 6: Explain negative resistance region in tunnel diode.
Question 7: Draw and explain an integrator circuit.
Question 8: Why a photodiode is used in reverse biased condition.
Question 9: Write down the features of varactor diode.
Question 10: Draw and explain the characteristics of a Schottky diode.
Question 11: What is dark current of photodiode.

Chapter 3
Transistor Amplifier Frequency Response

Learning Objectives

- **Define Bipolar transistor and its application i.e. working as an amplifier.**
- **Discuss Eber's mole model with mathematical expression.**
- **Introduce Miller's theorem and it's use in amplifier.**
- **Analyze high and low frequency response of common emitter & common source amplifier configuration.**
- **Discuss h-parameters for transistors.**
- **Analyze CE, CC, CB amplifier configuration using hybrid model and define their respective h-parameters.**

3.1 Introduction

Amplifiers are widely used electronic circuits that have the properties of amplification, hence their names. Amplifiers produce gain of an electrical signal with respect to its frequency. As these amplifiers use resistor, inductor, or capacitor (RLC) networks within their design, there is an important relationship between the use of these reactive components and the circuit's frequency response characteristics. Frequency response of an electric or electronics circuit allows us to see exactly how the output gain (known as the magnitude response) and the phase (known as the phase response) change at a particular single frequency, or over a whole range of different frequencies from 0 Hz (DC) to many thousands of megahertz (MHz), depending upon the design characteristics of the circuit.

Generally, the frequency response analysis of a transistor amplifier system is shown by plotting its gain that is the size of its output signal to its input signal. The frequency response of a given frequency-dependent circuit can be displayed as a

G. S. Tomar and A. Bagwari, *Fundamentals of Electronic Devices and Circuits*,
Algorithms for Intelligent Systems, https://doi.org/10.1007/978-981-15-0267-5_3

graphical sketch of magnitude (gain) against frequency (f). In this chapter, we discuss the characteristics of an amplifier, how transistor works as an amplifier, frequency response, Miller's theorem, frequency responses of amplifiers, as well as various amplifiers, uses of coupling with their types, and hybrid parameters.

3.2 Bipolar Transistor as an Amplifier

An amplifier is used to increase the single level, i.e., the amplifier is used to get a large output signal from a small input signal. We will assume a sinusoidal signal at the input of amplifier, at which the output signal must remain sinusoidal in waveform same as that of the input. The only difference between the input and output signal is that the output signal has high amplitude level with out of phase signal then the input signal. To make the transistor work as an amplifier it is to be biased to operate in the active region i.e. Base emitter junction is forward biased, while base collector junction is Reverse Biased. A transistor may be used to perform a number of functions, but many transistors are used in amplifying electric signal.

Figure 3.1 shows NPN transistor connected in CB configuration. The emitter is input terminal, and the collector is output terminal. Figure 3.2 shows that the low

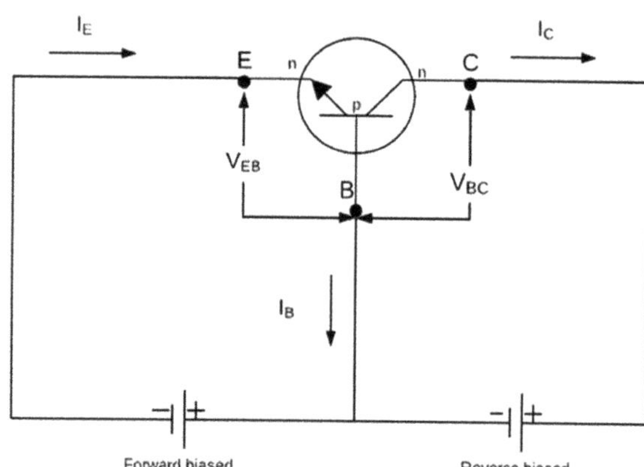

Fig. 3.1 Generalized CB configuration

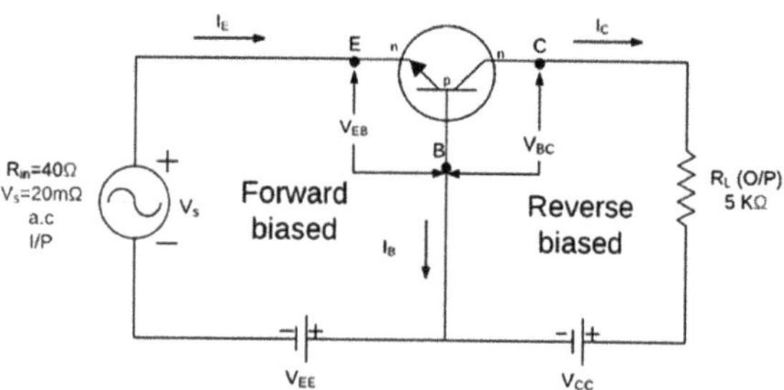

Fig. 3.2 CB configuration when low input signal is applied

input signal is applied between emitter base junction and the output is taken across load resistance R_L, which is connected in the collector circuit. As we know that transistor will work as an amplifier only if it operates in active region. In active region, V_{EE} is forward biased and V_{cc} is reverse biased.

Let us consider that an AC signal source V_s is connected in input circuit and load $R_L = 5\,k\Omega$ where $V_s = 20\,mV$ and input resistance $R_{in} = 40\,\Omega$. The input base junction is forward biased so it offers very low resistance to the input signal V_s ($R_{in} = 40\,\Omega$), while the output junction, B junction, is reverse biased so that output resistance is very high ($R_L = 5\,k\Omega$). The input current is:

$$I_E = \frac{V_{in}}{R_{in}} = \frac{20\,mV}{40} = 0.5\,mA.$$

As we know that $I_E = I_c + I_B$ and the value of I_B is negligible now, $I_E \cong I_c = 0.5\,mA$

The voltage across the load is:

$$V_0 = I_c \times R_c$$
$$V_0 = 5 \times 10^3 \times 0.5 \times 10^3$$
$$V_0 = 2.5\,V.$$

Now the voltage gain/voltage amplifier gain is:

$$A = \frac{\text{output}}{\text{input}} = \frac{V_0}{V_{in}} = \frac{2.5\,V}{20\,mV}$$
$$A = \frac{2.5 \times 10^3}{20}$$
$$A = 125$$
$$A = \frac{V_0}{V_{in}} = 125$$
$$V_0 = 125 V_{in}.$$

Output voltage is 125 times greater than the input voltage. This shows amplification.

3.3 Ebers–Moll Model

Ebers–Moll model was developed by Ebers and Moll in 1954. This model is also known as coupled diode model. This model involves two ideal current sources and two ideal diodes. As shown in figure, the PNP transistor is used in Ebers–Moll model. The diodes are placed back to back with reverse saturation current ($-I_{Eo}$ is reverse

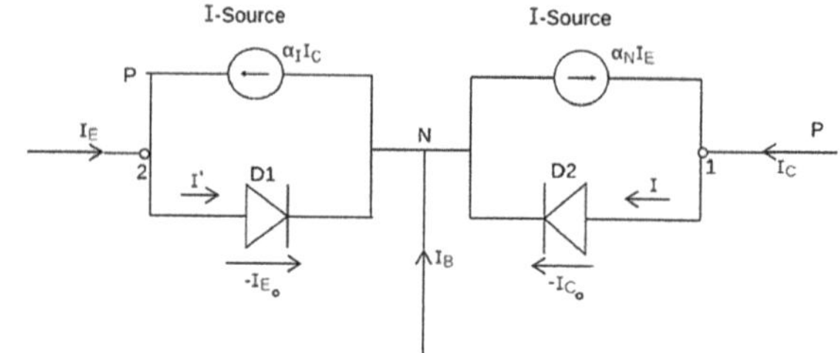

Fig. 3.3 Common base configuration (CB configuration) PNP configuration/coupled diode model

saturation emitter current and $-I_{\mathrm{Co}}$ is reverse saturation collector current), and two dependent current controlled current sources shunting the ideal diodes.

In the given Fig. 3.3, α_{I} is inverted common base current gain, α_{N} is normal common base I gain, I_{E} is emitter current base current gain, I_{C} is collector current base current gain, $-I_{\mathrm{Eo}}$ is reverse saturation emitter current, and $-I_{\mathrm{Co}}$ is reverse saturation collector current.

For a PNP transistor, both I_{Co} and I_{Eo} are negative, so that the magnitude of the reverse saturation current of the diodes becomes positive.

Let us see the equation I_{c} and I_{E} from Ebers–Moll model.

Applying KCL to the collector node at Point 1 in figure, now the mathematical expression becomes

$$\alpha_{\mathrm{N}} I_{\mathrm{E}} + I_{\mathrm{c}} - I = 0 \tag{3.1}$$

$$I_{\mathrm{c}} = -\alpha_{\mathrm{N}} I_{\mathrm{E}} + I \tag{3.2}$$

where

$$I = I_0\left(\mathrm{e}^{\frac{V_{\mathrm{c}}}{V_{\mathrm{T}}}} - 1\right) \text{ and } I_0 = -I_{\mathrm{Co}} \tag{3.3}$$

Then,

$$I = -I_{\mathrm{Co}}\left(\mathrm{e}^{\frac{V_{\mathrm{c}}}{V_{\mathrm{T}}}} - 1\right) \tag{3.4}$$

Now,

$$I_{\mathrm{C}} = -\alpha_{\mathrm{N}} I_{\mathrm{E}} - I_{\mathrm{Co}}\left(\mathrm{e}^{\frac{V_{\mathrm{c}}}{V_{\mathrm{T}}}} - 1\right) \tag{3.5}$$

Similarly, the emitter current will be at Point 2

$$I_{\mathrm{E}} = -\alpha_{\mathrm{I}} I_{\mathrm{c}} - I_{\mathrm{Co}}\left(\mathrm{e}^{\frac{V_{\mathrm{c}}}{V_{\mathrm{T}}}} - 1\right) \tag{3.6}$$

Now using Eqs. (3.5) and (3.6), putting the value of I_{E} into I_{c}

$$I_c = -\alpha_N\left\{-\alpha_I I_C - I_{Eo}\left(e^{\frac{V_c}{V_T}} - 1\right)\right\} - I_{Co}\left(e^{\frac{V_c}{V_T}} - 1\right) \tag{3.7}$$

$$I_c = \alpha_N\alpha_I I_c + \alpha_N I_{Eo}\left(e^{\frac{V_c}{V_T}} - 1\right) - I_{Co}\left(e^{\frac{V_c}{V_T}} - 1\right) \tag{3.8}$$

$$I_c - \alpha_N\alpha_I I_c = \alpha_N I_{Eo}\left(e^{\frac{V_c}{V_T}} - 1\right) - I_{Co}\left(e^{\frac{V_c}{V_T}} - 1\right) \tag{3.9}$$

$$I_c(1 - \alpha_I\alpha_N) = \alpha_N I_{Eo}\left(e^{\frac{V_c}{V_T}} - 1\right) - I_{Co}\left(e^{\frac{V_c}{V_T}} - 1\right) \tag{3.10}$$

$$I_c = \frac{\alpha_N I_{Eo}\left(e^{\frac{V_c}{V_T}} - 1\right)}{(1 - \alpha_I\alpha_N)} - \frac{I_{Co}\left(e^{\frac{V_c}{V_T}} - 1\right)}{(1 - \alpha_I\alpha_N)} \tag{3.11}$$

Now we can say that the value of I_E will be

$$I_E = \frac{\alpha_I I_{Co}\left(e^{\frac{V_c}{V_T}} - 1\right)}{(1 - \alpha_I\alpha_N)} - \frac{I_{Eo}\left(e^{\frac{V_c}{V_T}} - 1\right)}{(1 - \alpha_I\alpha_N)} \tag{3.12}$$

According to the physical principles,

$$I_c = a_{11}\left(e^{\frac{V_c}{V_T}} - 1\right) + a_{12}\left(e^{\frac{V_c}{V_T}} - 1\right) \tag{3.13}$$

$$I_E = a_{21}\left(e^{\frac{V_c}{V_T}} - 1\right) + a_{22}\left(e^{\frac{V_c}{V_T}} - 1\right) \tag{3.14}$$

If Eqs. (3.13) and (3.14) are satisfied, then

$$a_{11} = a_{21} \tag{3.15}$$

Now

$$a_{11} = \frac{\alpha_I I_{Co}}{(1 - \alpha_I\alpha_N)} \text{ and } a_{21} = \frac{\alpha_N I_{Eo}}{(1 - \alpha_I\alpha_N)} \tag{3.16}$$

Using Eq. (3.16), we can say that

$$\frac{\alpha_I I_{Co}}{(1 - \alpha_I\alpha_N)} = \frac{\alpha_N I_{Eo}}{(1 - \alpha_I\alpha_N)} \tag{3.17}$$

$$\alpha_I I_{Co} = \alpha_N I_{Eo} \tag{3.18}$$

$$\frac{\alpha_I}{\alpha_N} = \frac{I_{Eo}}{I_{Co}} \tag{3.19}$$

In Ebers–Moll model, voltage (V) is the function of current (I), as we know that the emitter current I_E is

$$I_{\mathrm{E}}=\frac{\alpha_{\mathrm{I}} I_{\mathrm{Co}}\left(\mathrm{e}^{\frac{V_{\mathrm{c}}}{V_{\mathrm{T}}}}-1\right)}{\left(1-\alpha_{\mathrm{I}} \alpha_{\mathrm{N}}\right)}-\frac{I_{\mathrm{Eo}}\left(\mathrm{e}^{\frac{V_{\mathrm{c}}}{V_{\mathrm{T}}}}-1\right)}{\left(1-\alpha_{\mathrm{I}} \alpha_{\mathrm{N}}\right)} \tag{3.20}$$

Now multiply by α_{N} in Eq. (3.20)

$$\alpha_{\mathrm{N}} I_{\mathrm{E}}=\frac{\alpha_{\mathrm{N}} \alpha_{\mathrm{I}} I_{\mathrm{Co}}\left(\mathrm{e}^{\frac{V_{\mathrm{c}}}{V_{\mathrm{T}}}}-1\right)}{\left(1-\alpha_{\mathrm{I}} \alpha_{\mathrm{N}}\right)}-\frac{\alpha_{\mathrm{N}} I_{\mathrm{Eo}}\left(\mathrm{e}^{\frac{V_{\mathrm{c}}}{V_{\mathrm{T}}}}-1\right)}{\left(1-\alpha_{\mathrm{I}} \alpha_{\mathrm{N}}\right)} \tag{3.21}$$

We know that the collector (I_{C}) is

$$I_{\mathrm{C}}=\frac{\alpha_{\mathrm{N}} I_{\mathrm{Eo}}\left(\mathrm{e}^{\frac{V_{\mathrm{c}}}{V_{\mathrm{T}}}}-1\right)}{\left(1-\alpha_{\mathrm{I}} \alpha_{\mathrm{N}}\right)}-\frac{I_{\mathrm{Co}}\left(\mathrm{e}^{\frac{V_{\mathrm{c}}}{V_{\mathrm{T}}}}-1\right)}{\left(1-\alpha_{\mathrm{I}} \alpha_{\mathrm{N}}\right)} \tag{3.22}$$

Adding Eqs. (3.21) and (3.22)

$$\begin{aligned} I_{\mathrm{C}}+\alpha_{\mathrm{N}} I_{\mathrm{E}}= & \frac{\alpha_{\mathrm{N}} \alpha_{\mathrm{I}} I_{\mathrm{Co}}\left(\mathrm{e}^{\frac{V_{\mathrm{c}}}{V_{\mathrm{T}}}}-1\right)}{\left(1-\alpha_{\mathrm{I}} \alpha_{\mathrm{N}}\right)}-\frac{\alpha_{\mathrm{N}} I_{\mathrm{Eo}}\left(\mathrm{e}^{\frac{V_{\mathrm{c}}}{V_{\mathrm{T}}}}-1\right)}{\left(1-\alpha_{\mathrm{I}} \alpha_{\mathrm{N}}\right)}+\frac{\alpha_{\mathrm{N}} I_{\mathrm{Eo}}\left(\mathrm{e}^{\frac{V_{\mathrm{c}}}{V_{\mathrm{T}}}}-1\right)}{\left(1-\alpha_{\mathrm{I}} \alpha_{\mathrm{N}}\right)} \\ & -\frac{I_{\mathrm{Co}}\left(\mathrm{e}^{\frac{V_{\mathrm{c}}}{V_{\mathrm{T}}}}-1\right)}{\left(1-\alpha_{\mathrm{I}} \alpha_{\mathrm{N}}\right)} \end{aligned} \tag{3.23}$$

$$I_{\mathrm{C}}+\alpha_{\mathrm{N}} I_{\mathrm{E}}=\frac{\alpha_{\mathrm{N}} \alpha_{\mathrm{I}} I_{\mathrm{Co}}\left(\mathrm{e}^{\frac{V_{\mathrm{c}}}{V_{\mathrm{T}}}}-1\right)}{\left(1-\alpha_{\mathrm{I}} \alpha_{\mathrm{N}}\right)}-\frac{I_{\mathrm{Co}}\left(\mathrm{e}^{\frac{V_{\mathrm{c}}}{V_{\mathrm{T}}}}-1\right)}{\left(1-\alpha_{\mathrm{I}} \alpha_{\mathrm{N}}\right)} \tag{3.24}$$

$$I_{\mathrm{C}}+\alpha_{\mathrm{N}} I_{\mathrm{E}}=\left(\frac{\left(\mathrm{e}^{\frac{V_{\mathrm{c}}}{V_{\mathrm{T}}}}-1\right)}{\left(1-\alpha_{\mathrm{I}} \alpha_{\mathrm{N}}\right)}\right) I_{\mathrm{Co}}\left[\alpha_{\mathrm{I}} \alpha_{\mathrm{N}}-1\right] \tag{3.25}$$

$$I_{\mathrm{C}}+\alpha_{\mathrm{N}} I_{\mathrm{E}}=-\frac{\left(\mathrm{e}^{\frac{V_{\mathrm{c}}}{V_{\mathrm{T}}}}-1\right)}{\left(1-\alpha_{\mathrm{I}} \alpha_{\mathrm{N}}\right)} I_{\mathrm{Co}}\left[\alpha_{\mathrm{I}} \alpha_{\mathrm{N}}-1\right] \tag{3.26}$$

$$I_{\mathrm{C}}+\alpha_{\mathrm{N}} I_{\mathrm{E}}=\left[\mathrm{e}^{\frac{V_{\mathrm{c}}}{V_{\mathrm{T}}}}-1\right] I_{\mathrm{Co}} \tag{3.27}$$

$$\frac{I_{\mathrm{C}}+\alpha_{\mathrm{N}} I_{\mathrm{E}}}{I_{\mathrm{Co}}}=1-\mathrm{e}^{\frac{V_{\mathrm{c}}}{V_{\mathrm{T}}}} \tag{3.28}$$

$$\mathrm{e}^{\frac{V_{\mathrm{c}}}{V_{\mathrm{T}}}}=1-\frac{I_{\mathrm{C}}+\alpha_{\mathrm{N}} I_{\mathrm{E}}}{I_{\mathrm{Co}}} \tag{3.29}$$

Taking log in both sides

$$\log_{e} e^{\frac{V_{\mathrm{c}}}{V_{\mathrm{T}}}}=\log_{e}\left[1-\left(\frac{I_{\mathrm{C}}+\alpha_{\mathrm{N}} I_{\mathrm{E}}}{I_{\mathrm{Co}}}\right)\right] \tag{3.30}$$

$$\left(\frac{V_C}{V_T}\right)\log_e e = \log_e\left[1 - \left(\frac{I_C + \alpha_N I_E}{I_{Co}}\right)\right] \tag{3.31}$$

$$\left(\frac{V_C}{V_T}\right) = \log_e\left[1 - \left(\frac{I_C + \alpha_N I_E}{I_{Co}}\right)\right] \tag{3.32}$$

Now the collector voltage will be

$$V_c = V_T \log_e\left[1 - \left(\frac{I_C + \alpha_N I_E}{I_{Co}}\right)\right] \tag{3.33}$$

Similarly

$$V_E = V_T \log_e\left[1 - \left(\frac{I_C + \alpha_N I_E}{I_{Co}}\right)\right] \tag{3.34}$$

Question 3.1: Determine the emitter voltage (V_E) for PNP transistor, if V_T (threshold voltage) $= 0.026$ V and $\alpha_I = 0.78$, $I_c = 20\,\text{mA}$, and $I_{EO} = 50\,\text{mA}$, $I_E = 10\,\text{mA}$.
Solution: The mathematical formula to calculate emitter voltage is given as

$$V_E = V_T \times \log\left[1 - \frac{I_E + \alpha_I I_c}{I_{EO}}\right]$$

$$V_E = 0.026 \times \log\left[1 - \frac{10 \times 10^{-3} + 0.78 \times 20 \times 10^{-3}}{50 \times 10^{-3}}\right]$$

$$V_E = 0.026 \times \log_e[1 - 0.512]$$

$$V_E = 0.026 \times \log_e(0.488)$$

$$V_E = 0.026 \times (-0.717439)$$

$$V_E = -0.018653\ \text{V}$$

$$V_E = -18.653\ \text{mV}$$

3.4 Miller's Theorem

For the analysis purpose in transistor amplifier, it is necessary to split the capacitor between input (base/gate) and the output (collector/drain); this can be achieved using Miller's theorem.

As shown in Fig. 3.4, A_V represents absolute voltage gain of the amplifier at midrange frequencies and "C" represents either C_{bc} or C_{gd} in case of BJT and FET, respectively.

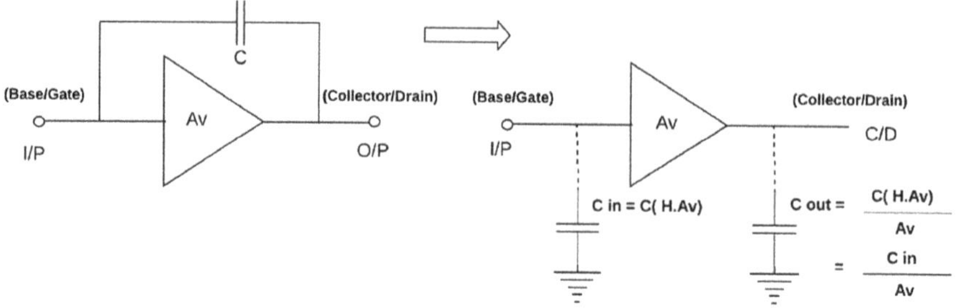

Fig. 3.4 Splitting of capacitor "*C*" using miller's Theorem

3.5 Low- and High-Frequency Response of Common Emitter Amplifier

3.5.1 High-Frequency Response of Common Emitter Amplifier

Figure 3.5a shows the circuit diagram of common emitter amplifier at high frequency. In this circuit, we have three capacitors, named as C_1 (direct capacitor), C_c (coupling capacitor), and C_E (bypass Capacitor). When we apply high frequency, C_c, C_E, and C_1 acts as a short circuit and do not affect the amplifier frequency response as shown in Fig. 3.5b. Now in Fig. 3.5c, all the capacitor acts a short circuit, and C_E and R_E show a short circuit.

We further find the resultant of R_1 and R_2 which is $R_1 \parallel R_2$ as shown in Fig. 3.5d. Now Fig. 3.5e shows high-frequency AC equivalent circuit, where $R_1 \parallel R_2$ and at the output $R_C \parallel R_L$. However, at higher frequencies, the internal capacitance plays an important role in case of high frequency for the given amplifier circuit.

Using Miller's theorem, the high-frequency AC equivalent circuit can be simplified as above. Now we are applying Miller's theorem to simplify the given circuit.

The internal capacitance C_{BC} can be split up into C_{in} (miller) and C_{out} (miller), where

$$C_{in}(\text{miller}) = C_{BC}(1 + A_V), \tag{3.35}$$

$$\text{And, } C_{out}(\text{miller}) = C_{BC}\left(\frac{1 + A_V}{A_V}\right), \tag{3.36}$$

As shown in Fig. 3.5h, there are two RC networks which affect the high-frequency response of the amplifier; these are

1. Input RC network.
2. Output RC network.

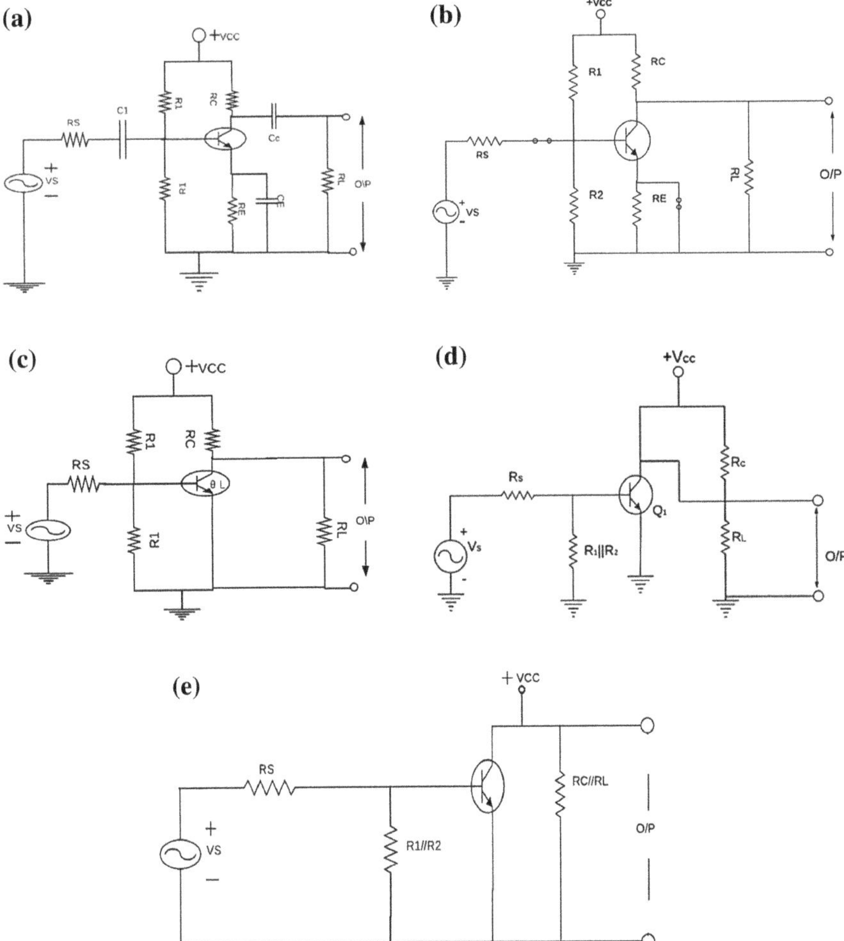

Fig. 3.5 **a** Common emitter. **b** "C" becomes short circuit amplifier at high frequency. **c** Short circuit. **d** Resultant Resistance. **e** High-frequency AC equivalent circuit. **f** Miller's theorem applied in Fig. 3.5e. **g** Miller's theorem output. **h** Simplified form of Fig. 3.5a

1. **Input RC Network**

This network is further reduced as shown in Fig. 3.6a. In high frequency, capacitive reactance becomes smaller. If we apply voltage divider theorem, then voltage is reduced with increase in frequency above midrange. This reduction in the signal reduces the circuit gain and hence output voltage. The f_C (cut-off frequency) can be calculated at C_{out} capacitive reactance which is equal to resistor $R_{i/p}$ as shown in Fig. 3.6b, i.e.,

$R_{i/p} = X_{C1} = R_S \parallel R_1 \parallel R_2 \parallel h_{ie}$

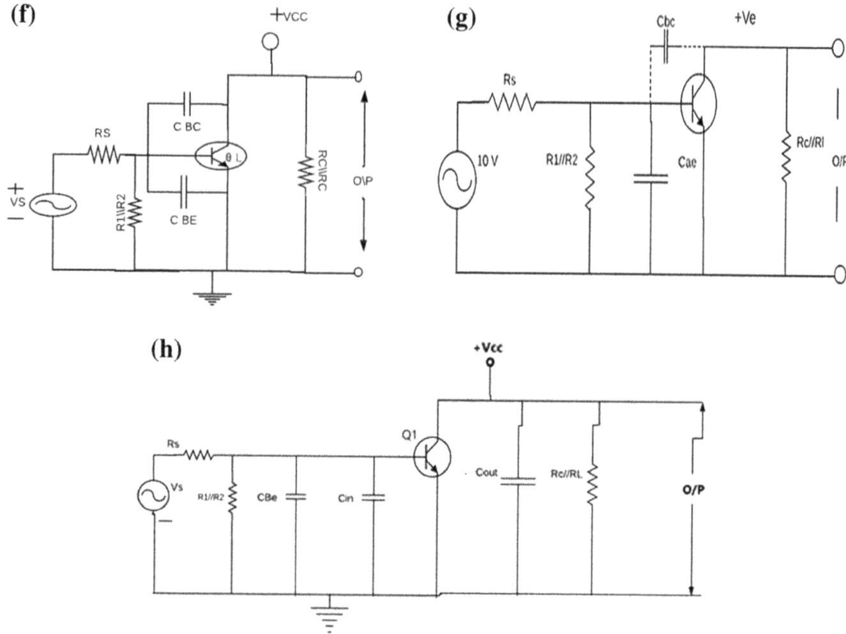

Fig. 3.5 (continued)

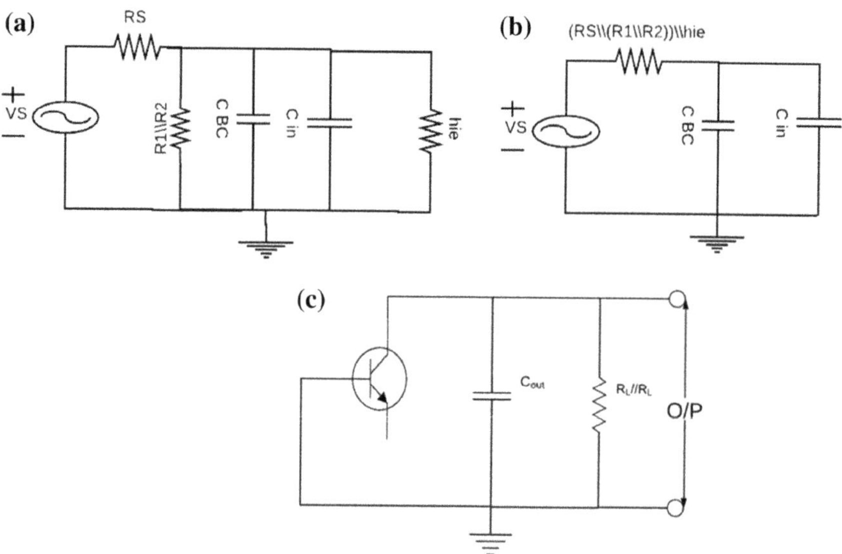

Fig. 3.6 **a** Input RC network. **b** Simplified form of Fig. 3.6a. **c** Output RC network

$$\text{and,} \quad f_{C(i/p)} = \frac{1}{2\pi R_{i/p} C_{i/p}} \tag{3.37}$$

where $C_{i/p} = (C_{BE} + C_{in})$, and then the overall cut-off frequency will be

$$f_{C(i/p)} = \frac{1}{2\pi [R_s \| (R_1 \| R_2) \| h_{ie}](C_{BE} + C_{in})} \tag{3.38}$$

2. **Output RC Network (Output Frequency Response)**

As we know that both networks have cut-off frequencies, then it is not necessary that these frequencies should be equal. Now, the network which has lower cut-off frequency than other network is known as dominant network.

Considering Fig. 3.6c, the frequency response of output

$$f_{C(o/p)} = \frac{1}{2\pi R_{o/p} C_{o/p}} \tag{3.39}$$

$$f_{C(o/p)} = \frac{1}{2\pi (R_c || R_L) C_{out(miller)}} \tag{3.40}$$

where

$$C_{in(miller)} = (A_V + 1)C_{BC}, \text{ and } C_{out(miller)} = \frac{(A_V + 1)C_{BC}}{A_V} \Rightarrow \frac{C_{in(miller)}}{A_V} \tag{3.41}$$

$$A_V = -h_{fe} \frac{R_{out}}{R_{in}} \tag{3.42}$$

where

$$R_{in} = h_{ie} \| R_1 \| R_2, \text{ and } R_{out} = R_C \| R_L \tag{3.43}$$

Using Eqs. (3.42) and (3.43), we get

$$A_V = \frac{-h_{fe}[R_C \| R_L]}{(R_1 \| R_2 \| h_{ie})} \tag{3.44}$$

We have seen that both the networks have cut-off frequencies. It is not necessary that these frequencies should be equal. The network which has lower cut-off frequency than other network is called dominant network.

Now, the phase shift in high-frequency output RC network is

$$\theta = \tan^{-1}\left(\frac{R_O}{X_{Cout(miller)}}\right) \tag{3.45}$$

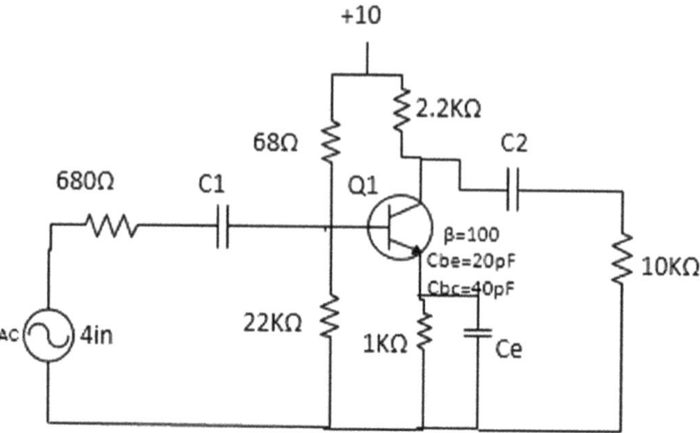

Fig. 3.7 RC network

Question 3.4: Determine the high-frequency response of the amplifier circuit shown in Fig. 3.7.

Solution: $A_V = -\frac{h_{fe} R_{out}}{R_{in}}$

$$R_{in} = h_{ie} \| R_1 \| R_2$$

$$R_{out} = R_C \| R_L$$

$$\text{and } A_V = \frac{-h_{fe}(R_C \| R_L)}{(h_{ie} \| R_1 \| R_2)} = \frac{-100(2.2\,\text{k}\Omega \| 10\,\text{k}\Omega)}{(1100 \| 68\,\text{k}\Omega \| 22\,\text{k}\Omega)} = \frac{-100(1.8\,\text{k}\Omega)}{1.032\,\text{k}\Omega} \Rightarrow -174.4,$$

Negative sign shows 180° phase shift between input and output,

$$C_{in}(\text{miller}) = C_{bc}(A_V + 1) \Rightarrow 4\,\text{pF}(174.4 + 1) \Rightarrow 0.7016\,\text{nF}.$$

$$C_{out}(\text{miller}) = \frac{C_{bc}(A_V + 1)}{(A_V)} \Rightarrow \frac{4\,\text{pF}(174.4 + 1)}{(174.4)} \Rightarrow 4\,\text{pF}.$$

(a)

$$f_C(\text{i/p}) = \frac{1}{2\pi \times [R_s \| R_1 \| R_2 \| h_{ie}][C_{be} + C_{in}(\text{miller})]} = \frac{1}{2\pi \times [(680 \| 68\,\text{k}\Omega) \| 22\,\text{k}\Omega \| 1100] \times [20\text{PF} + 0.7016\,\text{nF}]} = 537.947\,\text{KHz}.$$

(b)

$$f_C(\text{o/p}) = \frac{1}{2\pi \times (R_C \| R_L) \times C_{out}(\text{miller})} = \frac{1}{2\pi \times (2.2\,\text{k}\Omega \| 10\,\text{k}\Omega) \times 4\,\text{pF}} = 22.1\,\text{MHz}.$$

3.5.2 Low-Frequency Response of Common Emitter Amplifier

Let us consider a typical common emitter amplifier as shown in Fig. 3.8a. The amplifier shown in Fig. 3.8b has three RC networks that affect its gain as the frequency is decreased below midrange.

These are:

1. RC network formed by the input clamped capacitor C_1 and the input impedance of the amplifier.
2. RC network formed by the output coupling capacitor C_2. The resistance looking in at the collector and the load resistor R_L.
3. RC network formed by emitter bypass capacitor C_E and the resistance looking in at the emitter.

1. **Input RC Network**

Figure 3.8b shows input RC network formed by C_1 and input impedance (Z_{in}) of the amplifier. Note that V_{out} shown in Fig. 3.8b is the V_{out} of the network.

Applying "voltage divider theorem," we can write

$$V_{out} = \left[\frac{R_{in}}{\sqrt{(R_{in}^2 + X_{C1}^2)}}\right] \times V_{in} \tag{3.46}$$

We know that a critical point in the amplitude response is generally accepted to occur when the output voltage is 70.7% of the input.

$$V_{out} = 0.707 \times V_{in} \tag{3.47}$$

Thus, using Eqs. (3.46) and (3.47), we can write, at critical point,

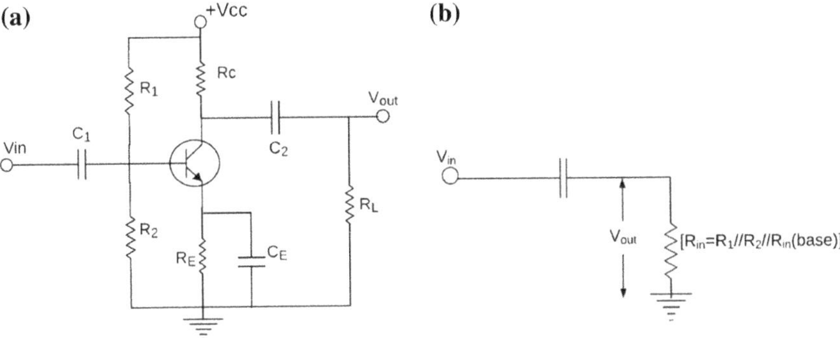

Fig. 3.8 **a** Typical RC coupled common emitter amplifier. **b** Input RC network

$$\frac{R_{\text{in}}}{\sqrt{(R_{\text{in}}^2 + X_{\text{C1}}^2)}} = 0.707 = \frac{1}{\sqrt{2}}$$

At this condition, $R_{\text{in}} = X_{\text{C1}}$

At this condition, the overall gain is reduced due to the attenuation provided by the input RC network.

The reduction in overall gain is given by

$$A_{\text{v}} = 20 \times \log_{10}\left(\frac{V_{\text{out}}}{V_{\text{in}}}\right) \Rightarrow 20 \times \log_{10}(0.707)$$
$$\cong -3\,\text{db}. \tag{3.48}$$

The f_{C} at this condition is called "lower critical frequency" and is given by

$$f_{\text{C}} = \frac{1}{2\pi R_{\text{in}} C_1} \tag{3.49}$$

where $R_{\text{in}} = R_1 \| R_2 \| h_{\text{ie}}$

The "R" input source is taken into account in the above equation because

$$f_{\text{C}(\text{i/p})} = \frac{1}{2\pi (R_{\text{S}} + R_{\text{in}}) C_1} \tag{3.50}$$

where $R_{\text{in}} = R_1 \| R_2 \| h_{\text{ie}}$

The phase angle in an input RC circuit is expressed as

$$\theta = \tan^{-1}\left(\frac{X_{C_1}}{R_{\text{in}}}\right) \tag{3.51}$$

2. **Output RC Network**

Figure 3.9a shows that output RC network formed by capacitor C_2 is the resistance looking in at the collector and the load "R_{L}".

The cut-off frequency "f_{C}" for this RC network is given by

$$f_{\text{C}(\text{o/p})} = \frac{1}{2 \times \pi \times (R_{\text{C}} + R_{\text{L}}) \times C_2} \tag{3.52}$$

The phase angle in the output RC circuit is expressed by

$$\theta = \tan^{-1}\left(\frac{X_{C_2}}{R_{\text{C}} + R_{\text{L}}}\right) \tag{3.53}$$

3. **Bypass Network**

Figure 3.9b shows the RC network formed by emitter bypass capacitor "C_{E}" and the resistance is looking in at the emitter.

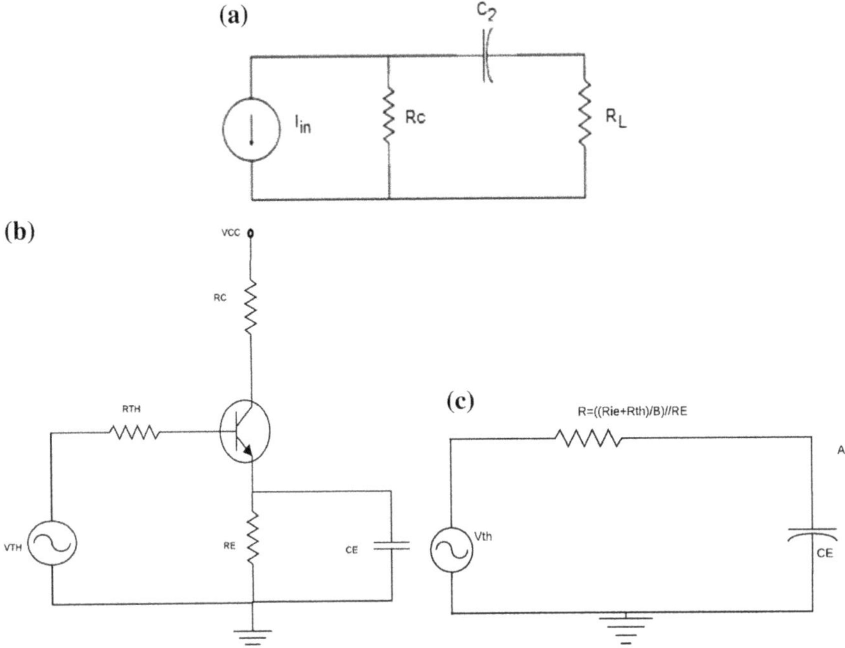

Fig. 3.9 **a** Output RC network. **b** Bypass RC network. **c** Simplified form of Fig. 3.9a

Here, $\left(\frac{h_{\mathrm{ie}}+R_{\mathrm{TH}}}{\beta}\right)$ is the resistance looking in at the emitter. It is derived as follows:

$$R = \left(\frac{V_{\mathrm{e}}}{I_{\mathrm{e}}} + \frac{h_{\mathrm{ie}}}{\beta}\right) \| R_{\mathrm{E}} \cong \left(\frac{V_{\mathrm{b}}}{\beta I_{\mathrm{b}}} + \frac{h_{\mathrm{ie}}}{\beta}\right) \| R_{\mathrm{E}}. \tag{3.54}$$

$$R \cong \left(\frac{I_{\mathrm{b}} R_{\mathrm{TH}}}{\beta I_{\mathrm{b}}} + \frac{h_{\mathrm{ie}}}{\beta}\right) \| R_{\mathrm{E}} \Rightarrow \left(\frac{R_{\mathrm{TH}} + h_{\mathrm{ie}}}{\beta}\right) \| R_{\mathrm{E}}. \tag{3.55}$$

Where,

$$R_{\mathrm{TH}} = (R_1 \| R_2 \| R_3), \tag{3.56}$$

(R_{TH} represents the Thevenin's equivalent resistance).
The cut-off frequency (f_{C}) for bypass network is

$$f_{\mathrm{C}} = \frac{1}{2\pi R C_{\mathrm{E}}} \tag{3.57}$$

Using Eqs. (3.55) and (3.57), we get

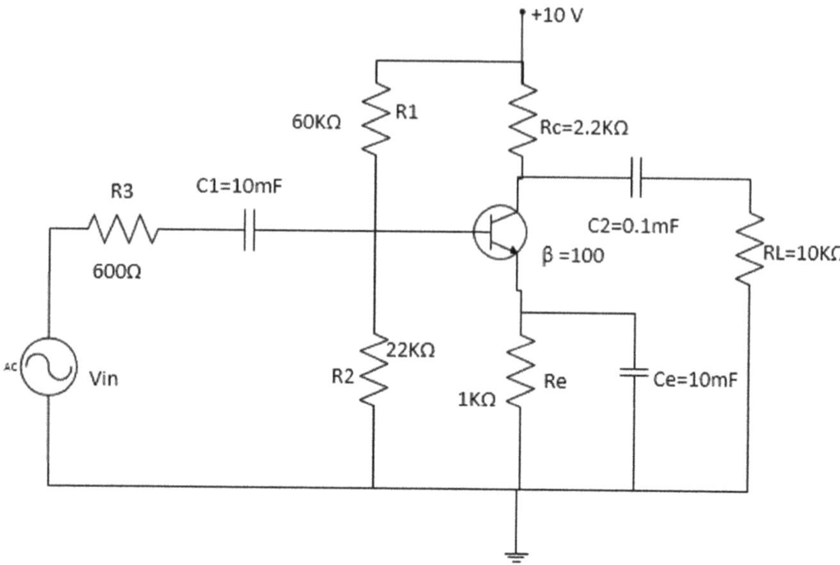

Fig. 3.10 RC network

$$f_{\mathrm{C}} = \frac{1}{2\pi\left[\left(\frac{h_{\mathrm{ie}}+R_{\mathrm{TH}}}{\beta}\right)\|R_{\mathrm{E}}\right]C_{\mathrm{E}}}, \tag{3.58}$$

The network which has higher cut-off frequency than other second network is called "dominant network." The dominant network determines the frequency at which the overall gain of the amplifier begins to drop at −20 dB/decade.

Question 3.5: Determine the low-frequency response of the amplifier circuit shown in Fig. 3.10, if h_{ie} is 1.1 kΩ.

Solution

(a) I/P RC network:

$$f_{\mathrm{C(i/p)}} = \frac{1}{2\pi[R_{\mathrm{s}}\|(R_1\|R_2)\|h_{\mathrm{ie}}](C_1)}$$

$$f_{\mathrm{C}}(\mathrm{i/p}) = \frac{1}{2\pi[680+(68\,\mathrm{k\Omega}||22\,\mathrm{k\Omega}||1.1\,\mathrm{k\Omega})]0.1\times10^{-6}} \Rightarrow 929.8\,\mathrm{Hz}.$$

(b) Output RC network:

$$f_{\mathrm{C(o/p)}} = \frac{1}{2\pi(R_{\mathrm{C}}+R_{\mathrm{L}})C_2} = \left[\frac{1}{2\pi(2.2\,\mathrm{k\Omega}+10\,\mathrm{k\Omega})0.1\times10^{-6}}\right] \Rightarrow 130.45\,\mathrm{Hz}.$$

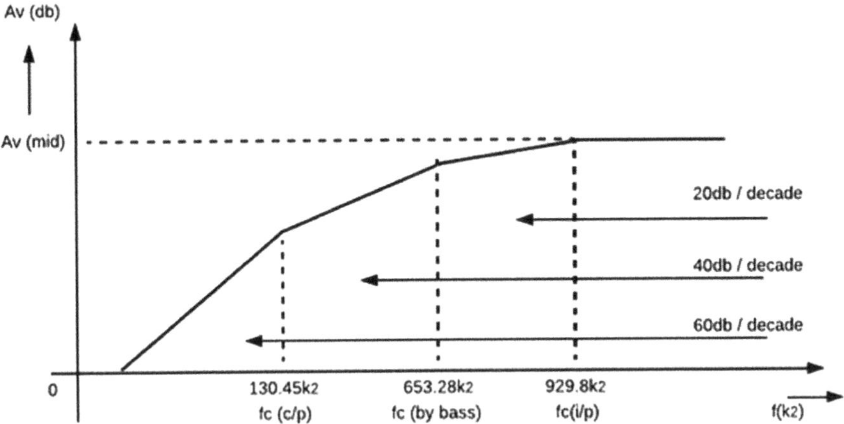

Fig. 3.11 Frequency response curve

(c) Bypass RC network:

$$f_C(\text{by pass}) = \frac{1}{2\pi\left[\left(\frac{R_{TH}+h_{ie}}{\beta}\right)\middle\|R_E\right]C_E} \Rightarrow \frac{1}{2\pi\left[\left(\frac{653.28+1100}{100}\right)\|1\,\text{k}\Omega\right]\times 10\times 10^{-6}}$$

$$R_{TH} = R_1\|R_2\|R_5 \Rightarrow (68\,\text{k}\Omega\|22\,\text{k}\Omega\|680) \Rightarrow 653.28\,\Omega.$$

$$f_C(\text{by pass}) = \frac{1}{2\pi(17.23)1010^{-6}} \Rightarrow 923.7\,\text{Hz}.$$

Considering input RC network, output RC network, and bypass network, the frequency response curve is drawn in Fig. 3.11.

3.6 Low- and High-Frequency Response of Common Source Amplifier

3.6.1 High-Frequency Response of Common Source Amplifier

In the given Fig. 3.12a at high frequencies, coupling and bypass capacitors act as short circuits and do not affect the amplifier high-frequency response. The equivalent circuit shows internal capacitance which affects the high-frequency response.

Using "Miller's theorem," this high-frequency equivalent circuit can be further simplified as follows in Fig. 3.12b.

The internal capacitance (C_{gd}) can be split into C_{in}(miller) and C_{out} (miller) as shown in Fig. 3.12c.

(a)

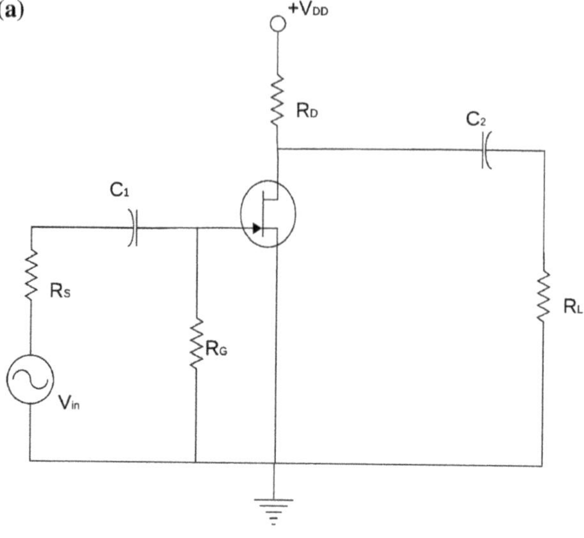

(b)

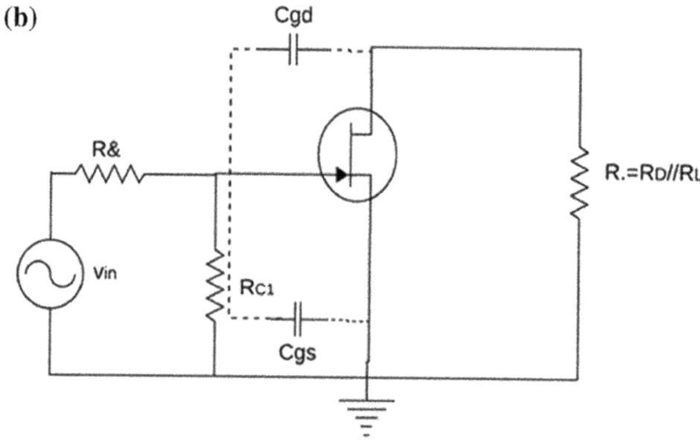

(c)

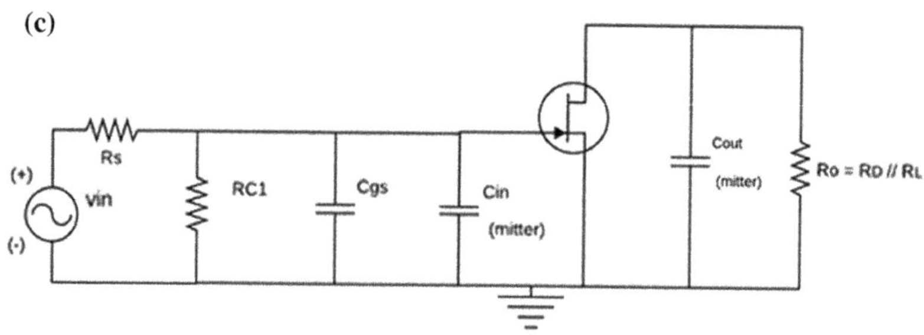

Fig. 3.12 **a** Typical RC coupled common source amplifier. **b** High-frequency equivalent circuit. **c** Simplified high-frequency equivalent circuit

where

$$C_{\text{in}}(\text{miller}) = C_{\text{gd}}(A_{\text{V}} + 1). \tag{3.59}$$

and,

$$C_{\text{out}}(\text{miller}) = C_{\text{gd}}\left(\frac{A_{\text{V}} + 1}{A_{\text{V}}}\right). \tag{3.60}$$

FET data sheets do not directly provide values for C_{gs} and C_{gd}. The data sheet normally provides values for input capacitance "C_{iss}" and the reverse transfer capacitance "C_{rss}". From C_{iss} and C_{rss}, the value for C_{gd} and C_{gs} can be calculated as follows:

$$C_{\text{gd}} = C_{\text{rss}} \tag{3.61}$$

$$C_{\text{gs}} = (C_{\text{iss}} - C_{\text{rss}}) \tag{3.62}$$

Figure 3.12c shows that there are two RC networks which affect the high-frequency response of the amplifier; these are

(1) Input RC network.
(2) Output RC network.

1. **Input RC Network**

Figure 3.13a shows input RC network. This network is further reduced as shown in Fig. 3.13b. Since ($R_{\text{S}} \ll R_{\text{G}}$)

$$f_{\text{C(i/p)}} = \frac{1}{2\pi R_{\text{S}} C_{\text{T}}} \tag{3.63}$$

As we know that,

$$C_{\text{T}} = C_{\text{gs}} + C_{\text{in(miller)}}. \tag{3.64}$$

Using Eqs. (3.63) and (3.64), we get

$$f_{\text{C(i/p)}} = \frac{1}{2\pi R_{\text{s}}\left[C_{\text{gs}} + C_{\text{in(miller)}}\right]} \tag{3.65}$$

The phase shift in high-frequency input RC network is

$$\theta = \tan^{-1}\left(\frac{R_{\text{s}}}{X_{\text{CT}}}\right). \tag{3.66}$$

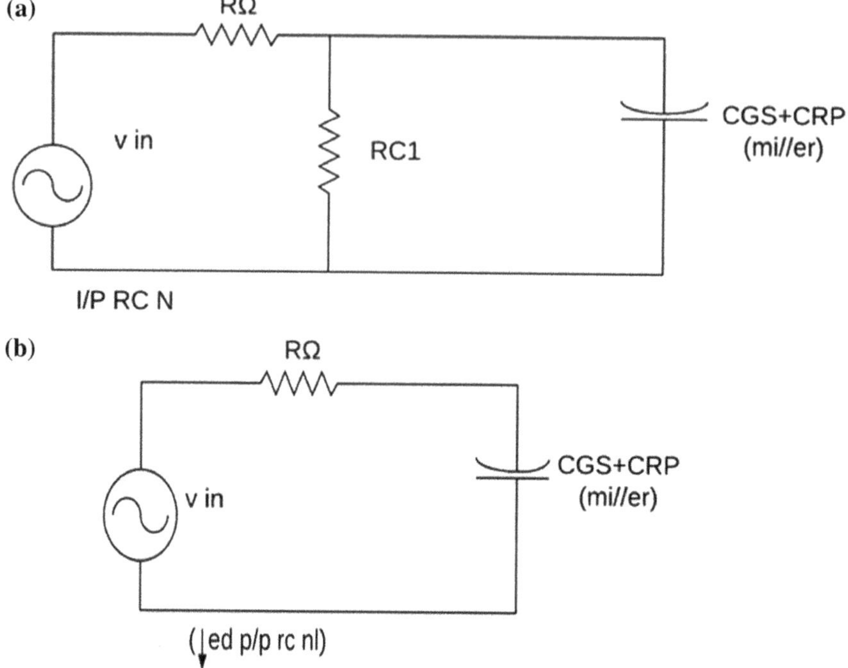

Fig. 3.13 **a** Input RC network. **b** Reduced input RC network

2. **Output RC Network**

Figure 3.14a shows the current source, circuit and Fig. 3.14b shows the voltage source circuit. Then the frequency response of the overall circuitry becomes

$$f_C = \frac{1}{2\pi\,(R_D \| R_L) C_{\text{out(miller)}}} \tag{3.67}$$

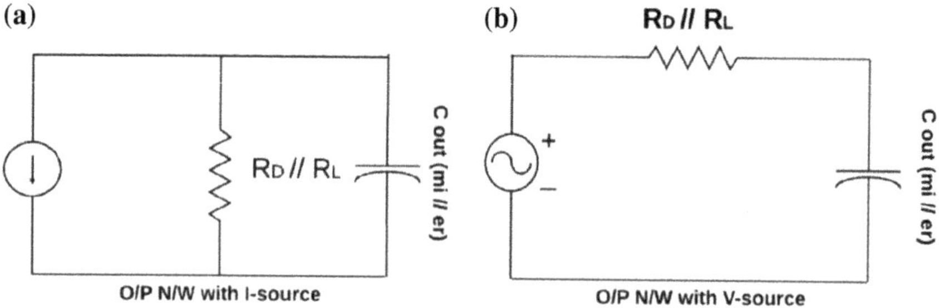

Fig. 3.14 **a** Output network with current source. **b** Output network with voltage source

Fig. 3.15 RC network

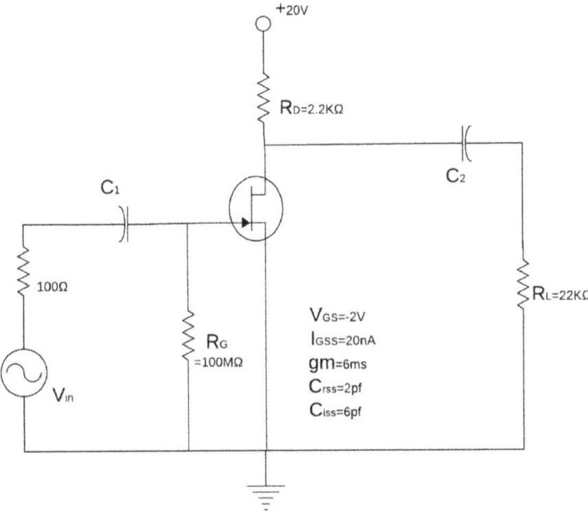

We have seen that both the networks have critical frequencies. It is not necessary that these frequencies should be "equal." The network which has lower cut-off frequency than other network is called "dominant network."

The phase shift in high-frequency output RC network is "θ"

$$\theta = \tan^{-1}\left(\frac{R_0}{X_{C_{\text{out(miller)}}}}\right). \tag{3.68}$$

Question 3.6: Determine the high-frequency response of the amplifier circuit as shown in Fig. 3.15.

Solution: $A_V = -g_m R_D$.

In the above equation, R_D should be replaced by ($R_D \parallel R_L$). Then

$$\begin{aligned} A_V &= -g_m(R_D||R_L) \\ &= -6\,\text{ms}(2.2\,\text{k}\Omega||22\,\text{k}\Omega) \\ &= -6\,\text{ms}(2\,\text{k}) \\ &= -12. \end{aligned}$$

$$\begin{aligned} C_{\text{in}}(\text{miller}) &= C_{\text{gd}}(A_V + 1) \\ &= C_{\text{rss}}(A_V + 1) \Rightarrow 2\,\text{pF}(12 + 1) \Rightarrow 26\,\text{pF}. \\ C_{\text{out}}(\text{miller}) &= C_{\text{gd}}\frac{(A_V + 1)}{A_V} = C_{\text{rss}}\frac{(A_V + 1)}{A_V} = \frac{2\text{pF}(12 + 1)}{(12)} = 2.166\,\text{pF}. \\ C_{\text{gs}} &= C_{\text{iss}} - C_{\text{rss}} \Rightarrow (6 - 2)\text{pF} \Rightarrow 4\,\text{Pf}. \end{aligned}$$

(a) $f_{C(\text{i/p})} = \frac{1}{2\pi R_S C_T} \Rightarrow \frac{1}{2\pi R_S[C_{\text{gs}} + C_{\text{in}}(\text{miller})]} \Rightarrow \frac{1}{2\pi(100)[4\,\text{pF} + 26\,\text{pF}]} \Rightarrow 53\,\text{MHz}.$

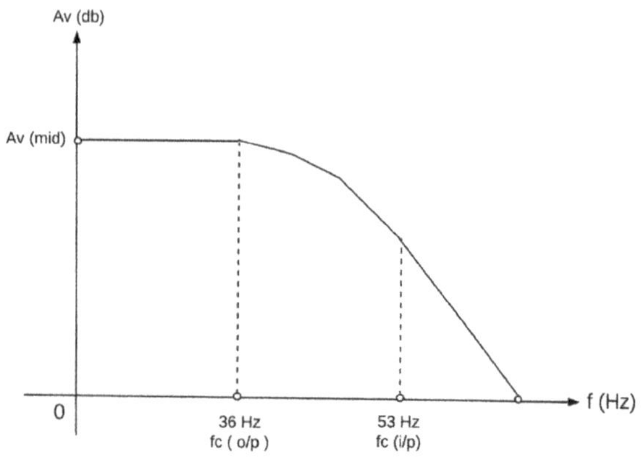

Fig. 3.16 Frequency response curve

(b) $f_{C(o/p)} = \frac{1}{2\pi(R_D \| R_L) \times C_{out\ (miller)}} \Rightarrow \frac{1}{2\pi[2.2\,k\Omega \| 22\,k\Omega] \times 2.166\,pF} \Rightarrow 36.74\,MHz.$

The value of the input frequency response and the output frequency response is considered, thus drawn the frequency response curve as shown in Fig. 3.16.

3.6.2 *Low-Frequency Response of Common Source Amplifier*

The amplifier shown in Fig. 3.17 have two RC networks that affect its gain as the frequency is reduced below midrange.

These are:

1. RC network formed by the input coupling capacitor C_1 and input impedance, and
2. RC network formed by the output coupling capacitor and the output impedance looking in at drain.

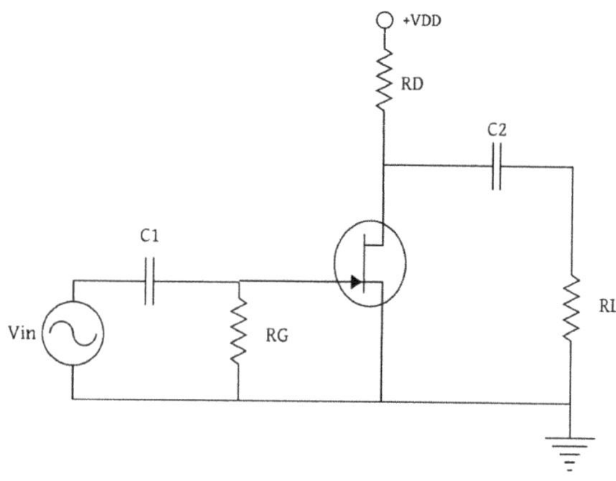

Fig. 3.17 Typical RC coupled common source amplifier

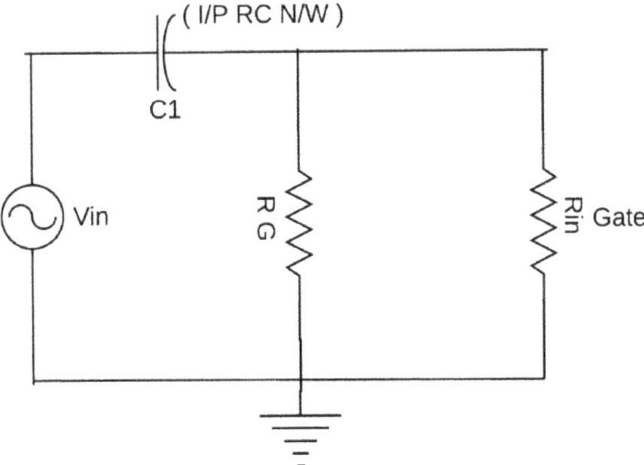

Fig. 3.18 Input RC network

1. **Input RC Network**

Figure 3.18 shows the input RC network formed by input impedance of the amplifier.

The low cut-off frequency of this network can be given as

$$f_C = \frac{1}{2\pi R_{in} C_1}. \tag{3.69}$$

where

$$R_{in} = R_C \| R_{in\,(Gate)} \tag{3.70}$$

and

$$R_{in(Gate)} = \left| \frac{V_{GS}}{I_{GSS}} \right| \tag{3.71}$$

where I_{GSS} indicates Gate reverse current.

Hence, the phase shift in the low-frequency RC circuit is

$$\theta = \tan^{-1}\left(\frac{X_{C1}}{R_{in}} \right). \tag{3.72}$$

2. **Output RC Network**

Figure 3.19 shows the output RC network formed by C_2 and the output impedance looking at the drain.

The lower cut-off frequency (f_C) for this network can be given as

$$f_C = \frac{1}{2\pi (R_D + R_L) C_2} \tag{3.73}$$

Fig. 3.19 Output RC network

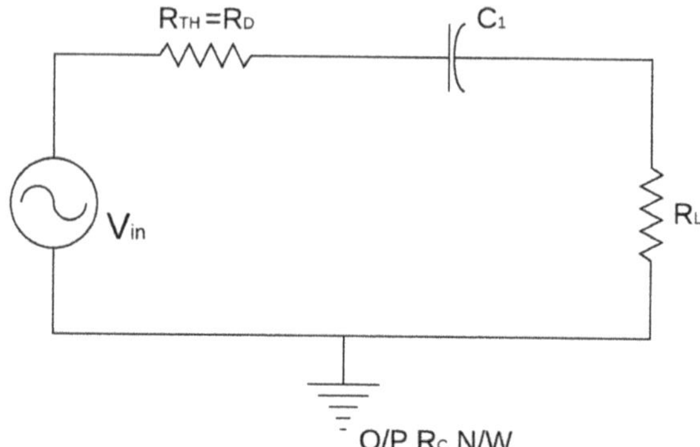

The FET amplifier circuit shown in Fig. 3.17 has two critical frequencies for two networks, and network having higher critical frequency is called "dominant network."

The phase shift in low-frequency output RC circuit is

$$\theta = \tan^{-1}\left(\frac{X_{C_2}}{R_D + R_L}\right). \tag{3.74}$$

Question 3.6: Determine the low f response of the amplifier circuit shown Fig. 3.20.
Solution: It is necessary to analyze each network to determine

(a) Input RC network,

$$fc = \frac{1}{2\pi R_{in} C_1}$$

$$R_{in} = R_G \| R_{in(Gate)} \Rightarrow R_G \| \left|\frac{V_{GS}}{I_{GSS}}\right|$$

$$R_{in} = 100\,\text{M}\Omega \| \frac{8}{8010^{-9}} \Rightarrow 100\,\text{M}\Omega| |100\,\text{M}\Omega \Rightarrow 50\,\text{M}\Omega.$$

$$f_c = \frac{1}{2\pi * 50\,\text{M}\Omega * 0.001 \times 10^{-6}} \Rightarrow 3.18\,\text{Hz}.$$

(b) Output RC network,

$$fc = \frac{1}{2\pi (R_D + R_L) C_2} \Rightarrow \frac{1}{2\pi (2.2\,\text{k}\Omega + 22\,\text{k}\Omega) 1 \times 10^{-6}} \Rightarrow 6.577\,\text{Hz}.$$

The value of the input RC network and the output RC network is considered, thus drawn the frequency response curve as shown in Fig. 3.21.

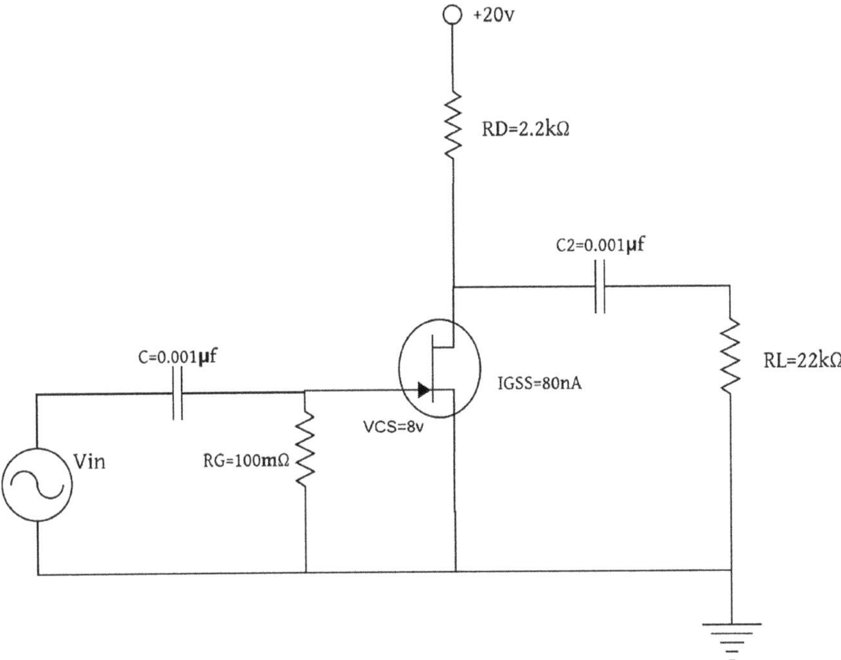

Fig. 3.20 RC network

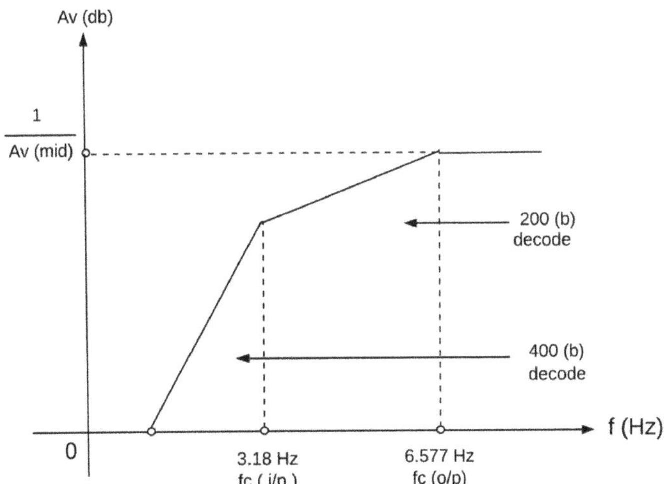

Fig. 3.21 Frequency response curve

3.7 Hybrid Parameters (h-Parameters)

As a matter of fact, if we carry out the analysis of the transistor amplifier, to study its behavior, with the help of β and the values of resistances used in the circuit, then this type of analysis is not found so accurate. In fact, the more accurate method to analyze the transistor amplifier is hybrid parameters or h-parameters method. We can study h-parameters at low frequency and at high frequency.

Hybrid parameters are also called as h-parameters. Characteristics of amplifier can be specified usually in terms of h-parameters. The hybrid or h-parameters are very popular and widely used because they give accurate result and can be measured very easily.

We know that the transistor is a three-terminal device. If one terminal is made common to input and output, then the transistor can be represented as a two-port network as shown in Fig. 3.22.

There are two ports, input port and output port. For each port, there are two variables, currents and voltages. Figure 3.22 shows the standard directions of currents and polarities of voltages for a two-port network. For a two-port network, the four variables are related in terms of h-parameters as

$$V_1 = h_{11}I_1 + h_{12}V_2 \tag{3.75}$$

$$I_2 = h_{21}I_1 + h_{22}V_2 \tag{3.76}$$

The parameters h_{11}, h_{12}, h_{21}, and h_{22} are called hybrid parameters or h-parameters.

3.7.1 Determination of h-Parameters

We can derive the expressions for different h-parameters as under:

(i) **Value of h_{11} and h_{21}**

If we short-circuit the output terminals, then $V_2 = 0$.

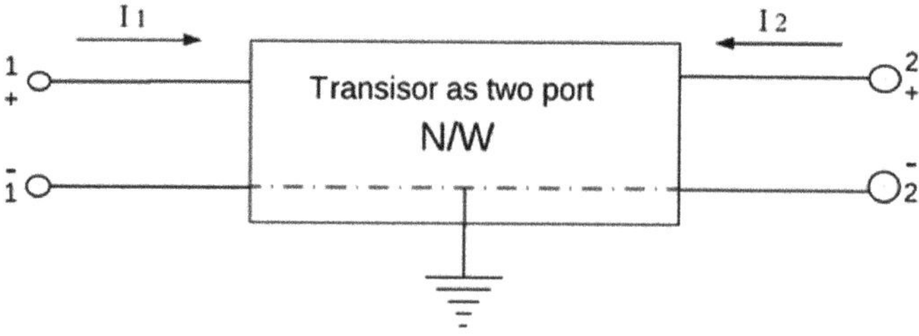

Fig. 3.22 Transistor as a two-port network

Putting this value $V_2 = 0$ in Eq. (3.75), we get

$$V_1 = h_{11}I_1 + h_{12}V_2 => h_{11}I_1 + 0$$
$$h_{11} = \frac{V_1}{I_1}$$

Since h_{11} is the ratio of input voltage and current, it is called **input impedance** with output short circuit.

Now putting $V_2 = 0$ in Eq. (3.76), we obtain $I_2 = h_{21}I_1 + h_{22}V_2 = h_{21}I_1 + 0$

$$h_{21} = \frac{I_2}{I_1}$$

Since h_{21} is the ratio of output and input currents, it is called **forward current gain** with output short circuit.

(ii) **Value of h_{12} and h_{22}**

If we open the input terminals, then $I_1 = 0$.
Putting $I_1 = 0$ in Eq. (3.75), we get

$$V_1 = h_{11}I_1 + h_{12}V_2 \Rightarrow 0 + h_{12}V_2$$
$$h_{12} = V_1/V_2$$

Since h_{12} is the ratio of input and output voltages, it is called **reverse voltage gain** with input open-circuited.

Now putting $I_1 = 0$ in Eq. (3.76), we get

$$I_2 = h_{21}I_1 + h_{22}V_2 \Rightarrow 0 + h_{22}V_2$$
$$h_{22} = I_2/V_2$$

Since h_{22} is the ratio of output current and output voltage, it is called **output admittance** with input open-circuited.

In short, h-parameters can be summarized as

1. $h_{11} = V_1/I_1 =$ input impedance with output shorted and unit is ohm.
   This is also represented as h_i

$$h_{11} = h_i.$$

2. $h_{21} = I_2/I_1 =$ forward current gain with output shorted and unit is none.
   This is also represented as h_f

$$h_{21} = h_f.$$

Table 3.1 h-Parameters and their meaning

S. No.	h-Parameter	Meaning	Condition
1	h_{11}	Input impedance	Output is shorted
2	h_{21}	Forward current gain	Output is shorted
3	h_{12}	Reverse voltage gain	Input open
4	h_{22}	Output admittance	Input open

3. $h_{12} = V_1/V_2$ = reverse voltage gain with input open-circuited and has no unit. This is also represented as h_r

$$h_{12} = h_r.$$

4. $h_{22} = I_2/V_2$ = output admittance with input open-circuited and has unit ohm^{-1} or mho. This is also represented as h_o.

$$h_{22} = h_o.$$

After analyzed above, it is clear that all the four parameters have different units, and thus, the parameters are hybrid (mixed) in nature. Therefore, they are called **hybrid or h-parameters** as shown in Table 3.1.

3.8 h-Parameter Representation for Transistor

The h-parameter representation for transistor depends upon the type of configuration (CE, CB, or CC) as shown in Table 3.2. Therefore, each h-parameter is assigned a second subscript letter *e*, *b*, or *c* depending upon the type of configuration. The letter "*e*" is used when the configuration is common emitter, the letter "*b*" is used if the configuration is common base, and the letter "*c*" is used if the configuration is common collector.

Table 3.2 h-Parameter representation for three types of configuration

S. No	h-Parameter	Types of configuration		
		Common emitter	Common base	Common collector
1	h_{11}	h_{ie}	h_{ib}	h_{ic}
2	h_{21}	h_{fe}	h_{fb}	h_{fc}
3	h_{12}	h_{re}	h_{rb}	h_{rc}
4	h_{22}	h_{oe}	h_{ob}	h_{oc}

3.9 General Hybrid Equivalent Circuit

There are three transistor configurations: CB, CE, and CC. Each configuration will have its separate hybrid or h-parameter equivalent circuit. But, first we develop a general h-parameter equivalent circuit which can be applied for any of the three configurations. The equivalent circuit thus obtained for any type of configuration is also called as small-signal, low-frequency hybrid model of a transistor.

Thus, h-parameter equations are

$$V_1 = h_{11}I_1 + h_{12}V_2$$
$$I_2 = h_{21}I_1 + h_{22}V_2$$

or

$$V_1 = h_i I_1 + h_r V_2 \quad (3.77)$$

$$I_2 = h_f I_1 + h_o V_2 \quad (3.78)$$

Since each term of Eq. (3.77) has the units of volts, we can use Kirchhoff's voltage law to find a circuit that fits this equation. Figure 3.23 shows the circuit.

Similarly, each term of Eq. (3.78) has the units of currents; we can use Kirchhoff's current law to find a circuit that fits this equation. Figure 3.24 shows this circuit.

Combining Figs. 3.23 and 3.24, we get the general hybrid (h-parameter) equivalent circuit as shown in Fig. 3.25.

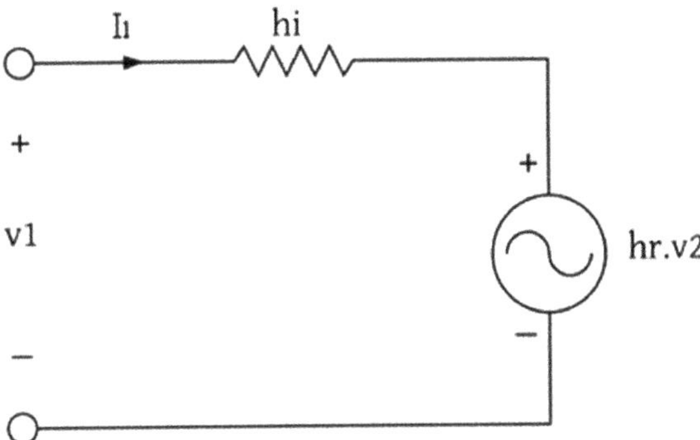

Fig. 3.23 Simple circuitry of low-frequency hybrid model of a transistor

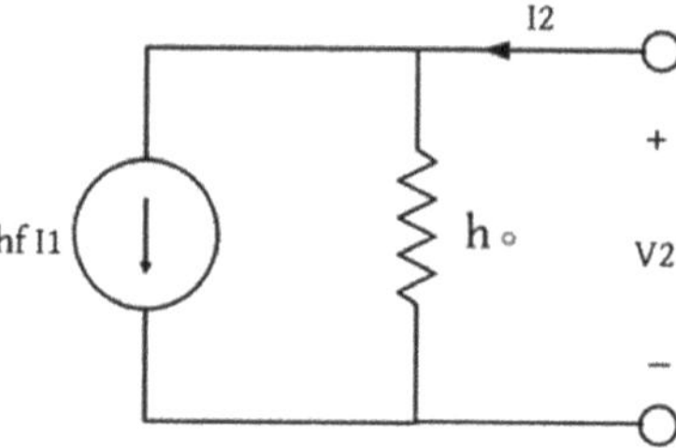

Fig. 3.24 Current source circuit

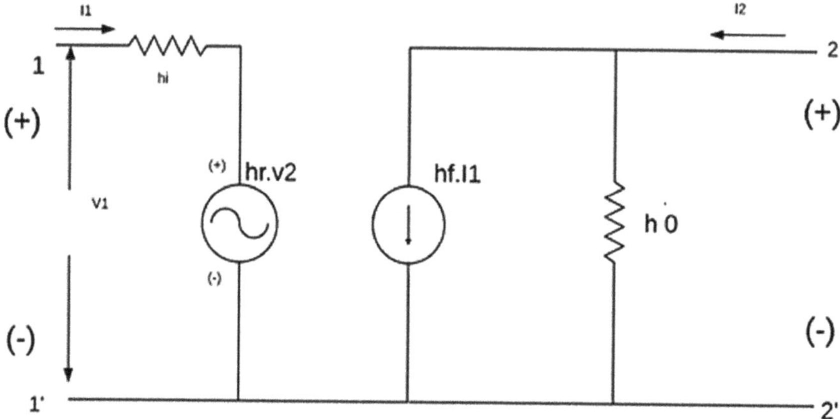

Fig. 3.25 General h-parameter equivalent circuit

3.10 Analysis Using Simplified Hybrid Model

So far we have seen the exact calculations of current gain, voltage gain, and input and output impedances of transistor amplifier circuits. In most practical cases, it is appropriate to obtain approximate values of current gain, voltage gain, and input and output impedances rather than to carry and put more lengthy exact calculations. We can justify this statement with the fact that h-parameter themselves usually vary widely for the same type of transistor. But now the question is when to use approximate analysis? To solve this question, there is a generalized rule; this rule says that if $h_{oe}R_L < 0.1$, then we can proceed for the approximation analysis, otherwise do the exact analysis. In this section, we see how to analyze transistor with approximate model.

3.10.1 *Analysis of Common Emitter Circuit Using Simplified Hybrid Model*

Let us consider the h-parameter equivalent circuit for the amplifier as shown in Fig. 3.26.

Now, see how can we modify this model so as to make the analysis simple without greatly sacrificing accuracy? Since $1/h_{oe}$ is in parallel with R_L and R_c if $1/h_{oe} \gg R_L \| R_C$, then h_{oe} may be neglected. If we neglect h_{oe}, the collector current I_C is given by $I_C = h_{fe} I_b$. Under these conditions, the magnitude of the voltage of the generator in the emitter circuit is

$$h_{re} \times |V_{ce}| = h_{re} \times I_C \times (R_1 \| R_C) = h_{re} \times h_{fe} \times I_b \times (R_L \| R_C)$$

Since $h_{re} h_{fe} \approx 0.01$, this voltage may be neglected in comparison with the $h_{ie} I_b$ drop across h_{ie}, provided that $R_L \| R_C$ is not too large. We, therefore, conclude that if the load resistance $R_L \| R_C$ is small, it is possible to neglect the parameters h_{re} and h_{oe} in the h-parameter equivalent circuit. Figure 3.27 shows the approximate h-parameter equivalent circuit.

(i) **Current Gain**: The common emitter (CE) current gain is given as

$$A_i = \frac{-I_c}{I_b} \Rightarrow \frac{-h_{fe}}{1 + h_{oe} R_L} \tag{3.79}$$

By neglecting h_{oe}, we have $A_i = -h_{fe}$

(ii) **Input Impedance**: The common emitter (CE) input impedance is given as

$$R_i = h_{ie} + h_{re} A_i R_L \tag{3.80}$$

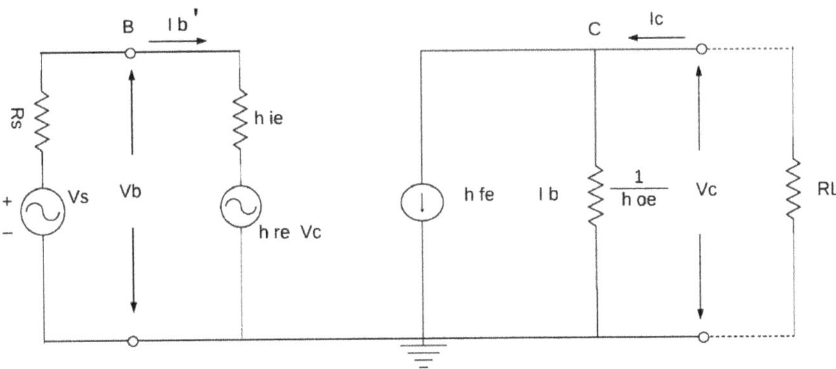

Fig. 3.26 h-Parameter equivalent circuit for the amplifier

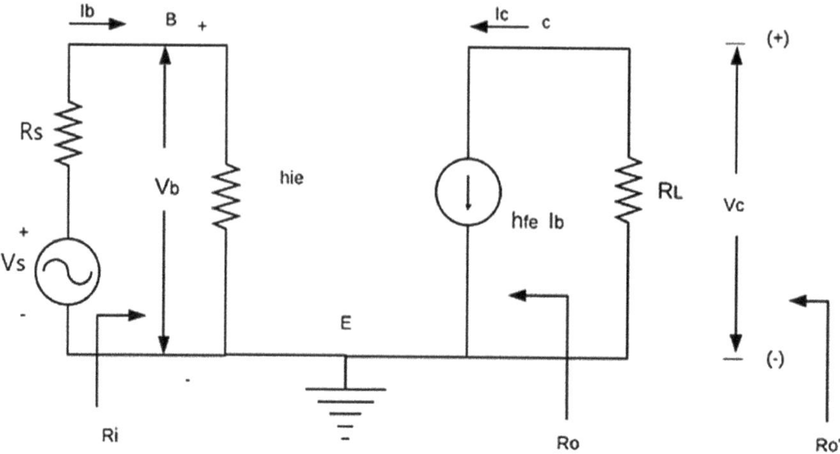

Fig. 3.27 Approximate common emitter model

By neglecting h_{re}, we have $R_i \approx h_{ie}$

(iii) **Voltage Gain**: The common emitter (CE) voltage gain is given as

$$A_v = \frac{A_i R_L}{R_i} \Rightarrow \frac{A_i R_L}{h_{ie}} \tag{3.81}$$

(iv) **Output Impedance**: The common emitter (CE) output impedance is given as

$$Y_o = h_{oe} - \frac{h_{fe} h_{re}}{h_{ie} + R_s}$$

By neglecting h_{oe} and h_{re}

$$Y_o = 0$$
$$R_o = \frac{1}{Y_o} = \infty \tag{3.82}$$

$$R_o^{'} = R_o \| R_L \Rightarrow \infty \| R_L \Rightarrow R_L \tag{3.83}$$

Example 3.4: Consider a single-stage CE amplifier with $R_s = 1\,\text{K}$, $R_1 = 50\,\text{K}$, $R_2 = 2\,\text{K}$, $R_C = 2\,\text{K}$, $R_L = 2\,\text{K}$, $h_{fe} = 50$, $h_{ie} = 1.1\,\text{K}$, $h_{oe} = 25\,\mu\text{A/V}$, and $h_{re} = 2.5 \times 10^{-4}$, as shown in Fig. 3.28.
Find A_i, R_i, A_v, and R_o.
Solution Since $h_{oe} R_L^{'} = 25 \times 10^{-6} \times 2\,\text{K} \| 2\,\text{K} \Rightarrow 0.025$, which is less than 0.1, we use approximate analysis.

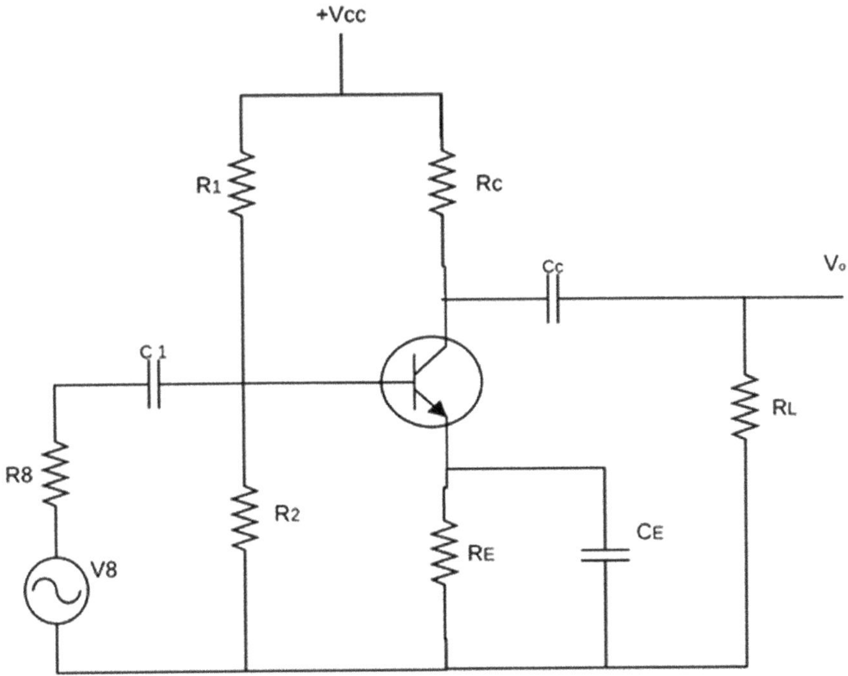

Fig. 3.28 Approximate common emitter model

Figure 3.29a shows the simplified hybrid model for the given circuit.

(a) **Current Gain**: $(A_i) = -h_{fe} \Rightarrow -50$
(b) **Input Impedance**: $(R_i) = h_{fe} \Rightarrow 1.1\,\text{K}$

$$R_i' = h_{ie} \| R_1 \| R_2 \Rightarrow 1.1\,\text{K} \| 50\,\text{K} \| 2\,\text{K} \Rightarrow 700\,\Omega.$$

(c) **Voltage Gain**: $(A_v) = \frac{A_i R_L}{R_i} \Rightarrow \frac{-50 \times (2\,\text{K} \| 2\,\text{K})}{1.1\,\text{K}} \Rightarrow -45.45,$
(d) **Output Impedance**: $(R_o) = \frac{1}{Y_o} = \infty$

$$R_o' = R_o \| R_L' \Rightarrow \infty \| 2\,\text{K} \| 2\,\text{K} \Rightarrow 1\,\text{K}\Omega.$$

(e) **Overall Voltage Gain**: $(A_{vs}) = \frac{V_o}{V_s} \Rightarrow \left(\frac{V_o}{V_b}\right) \times \left(\frac{V_b}{V_s}\right)$

where $\frac{V_o}{V_b} = A_v$, and $\frac{V_b}{V_s} = \frac{R_i'}{R_{i+R_s}'}$

$$A_{vs} = \frac{A_v R_i'}{R_i' + R_s} \Rightarrow \frac{-45.45 \times 700}{700 + 1\,\text{K}} \Rightarrow -18.71,$$

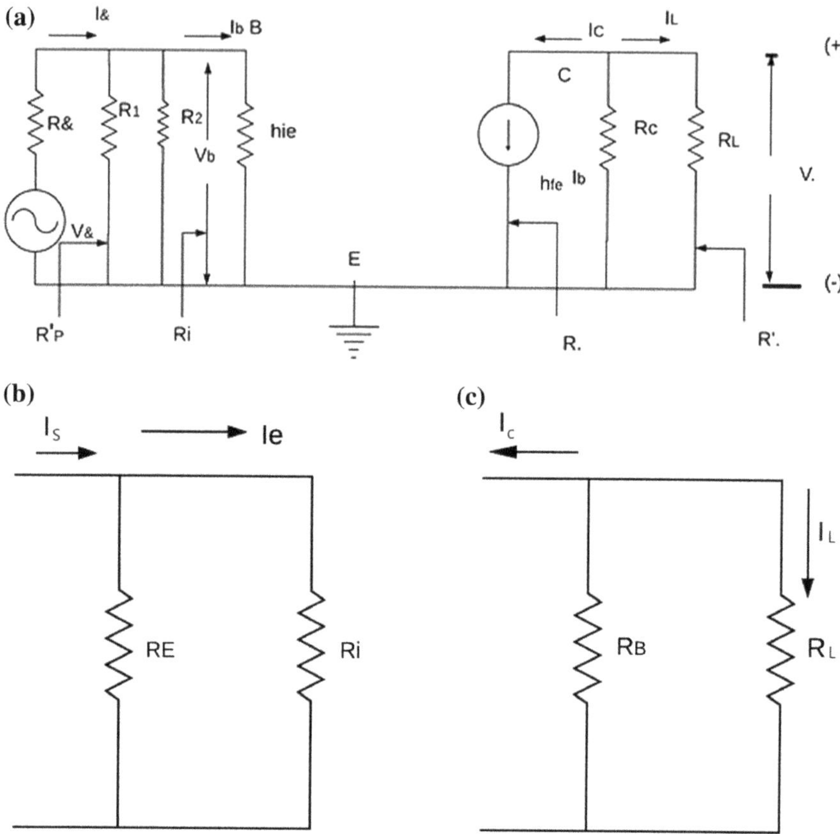

Fig. 3.29 **a** Simplified hybrid model. **b** Circuit using I_s and I_e. **c** Circuit using I_c and I_L

(f) **Overall Current Gain**: $A_i = \frac{I_L}{I_S} \Rightarrow \frac{I_L}{I_C} \times \frac{I_C}{I_B} \times \frac{I_B}{I_S}$

$$\frac{I_L}{I_C} = \frac{-R_C}{R_C + R_L} \Rightarrow \frac{-1\,K}{1\,K + 1\,K} \Rightarrow -0.5,$$

$$\frac{I_C}{I_B} = h_{fe} \Rightarrow 50.$$

Looking at Fig. 3.29c, $\frac{I_B}{I_S} = \frac{R_B}{R_B + R_i} \Rightarrow \frac{50\|2}{(50\|2)+1.1} \Rightarrow 0.636,$

$$A_{i\,(\text{for circuit})} = \frac{I_L}{I_s} \Rightarrow -0.5 \times 50 \times 0.636 \Rightarrow -15.9.$$

3.10.2 Analysis of Common Collector Circuit Using Simplified Hybrid Model

We have seen the simplified CE model, in which input is applied to base, and output is taken from collector, and emitter is common between input and output. The same simplified model can be modified to get simplified CC model. For simplified CC model, we have to make collector common and take the output from emitter, as shown in Fig. 3.30. The $h_{fe}I_b$ is now exactly opposite of CE model because the current $h_{fe}I_b$ always points toward emitter.

(i) **Current Gain**: It is defined as the ratio of output-to-input currents

$$A_i = \frac{I_o}{I_b} \Rightarrow \frac{-I_e}{I_b} \Rightarrow (1 + h_{fe}) \tag{3.84}$$

(ii) **Input Resistance**: From Fig. 3.30, we obtain $R_i = V_b/I_b$

Applying KVL, we have $V_b - I_b h_{ie} - I_o R_L = 0$

$$V_b = I_b h_{ie} + I_o R_L$$
$$\frac{V_b}{I_b} = h_{ie} + \frac{I_o}{I_b} \times R_L$$
$$R_i = \frac{V_b}{I_b} = h_{ie} + (1 + h_{fe}) \times R_L$$
$$\frac{I_o}{I_b} = -\frac{I_e}{I_b} = 1 + h_{fe}$$

The above equation shows that the input impedance of CC is higher than the CE configuration.

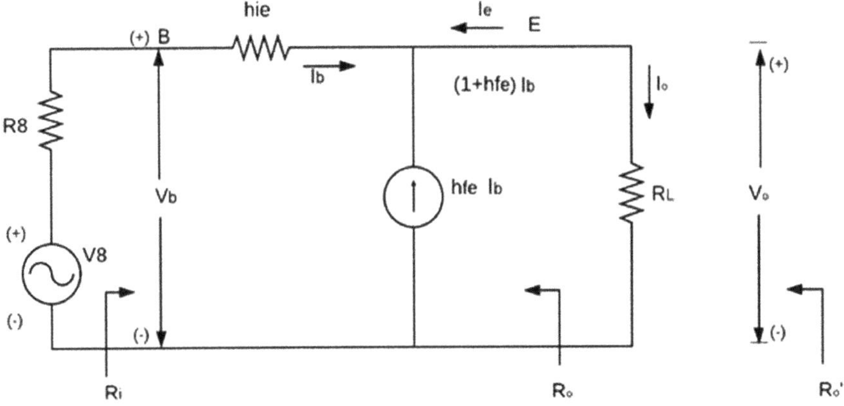

Fig. 3.30 Simplified common collector model

(iii) **Voltage Gain (A_v)**: It is given as $A_v = \frac{V_o}{V_b} \Rightarrow \frac{I_o R_L}{I_b R_i} \Rightarrow \frac{A_i R_L}{R_i}$

We know that $A_i = \frac{I_o}{I_b} \Rightarrow -\frac{I_e}{I_b}$
Substituting value of A_i and R_i, we get
$A_v = \frac{(1+h_{fe})R_L}{h_{ie}+(1+h_{fe})R_L} \cong 1$ but always less than $1, \because (1 + h_{fe})R_L \gg h_{ie}$

(iv) **Output Resistance (R_o)**:

It is the ratio of output voltage V_o to output current I_e with $V_s = 0$.

$$R_o = \frac{V_o}{I_e} \quad \text{when } V_s = 0$$

Applying KVL, we have

$$\begin{aligned}
&V_s - I_b R_s - I_b h_{ie} - V_o = 0 \\
&V_o = -I_b R_s - I_b h_{ie} \because V_o = 0 = -I_b(R_s + h_{ie}) \\
&I_e = -(1 + h_{fe}) I_b \\
&\frac{V_o}{I_e} = \frac{-I_b \times (R_s + h_{ie})}{-(1 + h_{fe}) \times I_b} \\
&R_o = \frac{V_o}{I_e} \\
&R_o = \frac{R_s + h_{ie}}{1 + h_{fe}}
\end{aligned}$$

The output resistance R_o' of the stage, taking the load into account, is given as

$$R_o' = R_o \| R_L$$

Example 3.5: A common collector circuit as shown in Fig. 3.31 has the following components: $R_1 = 27\,\text{K}\Omega$, $R_2 = 27\,\text{K}\Omega$, $R_E = 5.6\,\text{K}\Omega$, $R_L = 47\,\text{K}\Omega$, $R_s = 600\,\Omega$. The transistor parameters are $h_{ie} = 1\,\text{K}\Omega$, $h_{fe} = 85$, and $h_{oe} = 2\,\mu\text{A/V}$. Calculate $A_i, R_i, A_v, R_o, A_{vs}, A_i$.
Solution: Here, $h_{oe} \times R_L' = 2 \times 10^{-6} \times (5.6\|47\,\text{K}) \Rightarrow 0.01$, which is less than 0.1. Thus, we analyze the circuit with approximate model.

Figure 3.32a shows the simplified hybrid model of the given circuit.

(a) **Current Gain**: $(A_i) = (1 + h_{fe}) \Rightarrow 1 + 85 \Rightarrow 86.$
(b) **Input Resistance**:

$$\begin{aligned}
(R_i) &= h_{ie} + (1 + h_{fe}) \times R_L' \Rightarrow h_{ie} + (1 + h_{fe}) \times R_E \| R_L \\
&= 1\,\text{K} \times (1 + 85) \times (5.6\text{K}\|47\,\text{K}) \Rightarrow 431.33\,\text{K}\Omega. \\
R_i' &= R_i \| R_1 \| R_2 \Rightarrow 431.33\,\text{K}\|27\,\text{K}\|27\,\text{K}\Omega. \\
&= 13.09\,\text{K}\Omega.
\end{aligned}$$

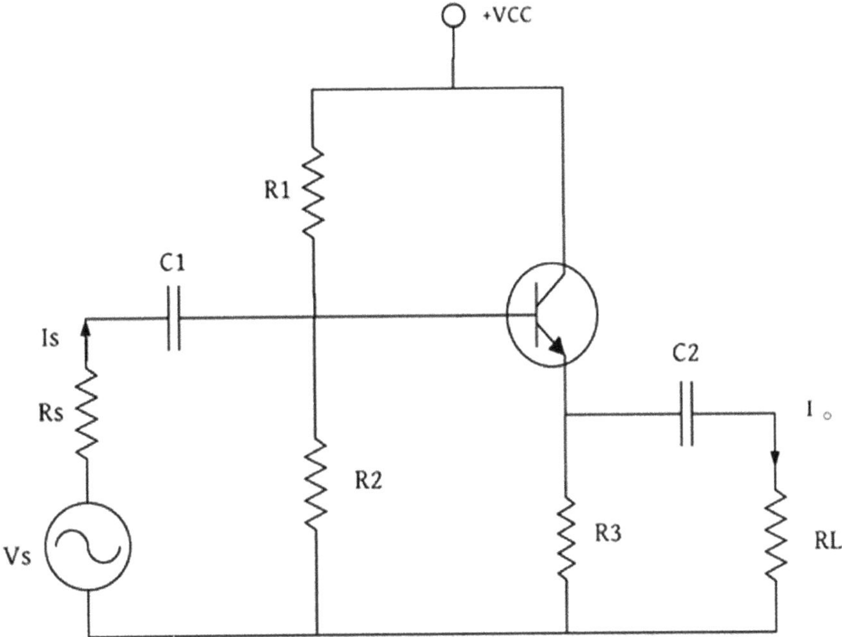

Fig. 3.31 Simplified common collector model

(c) **Voltage Gain**:

$$(A_v) = \frac{(1 + h_{fe}) \times R'_L}{h_{ie} + (1 + h_{fe}) \times R'_L}$$
$$= \frac{(1 + 85) \times (5.6\,\text{K}\|47\,\text{K}\Omega)}{1\,\text{K}\Omega + (1 + 85) \times (5.6\|47\,\text{K}\Omega)} \Rightarrow 0.997.$$

(d) **Output Resistance**:
Looking Fig. 3.32a the output resistance

$$R_o = \frac{R'_s + h_{ie}}{1 + h_{fe}} \Rightarrow \frac{(R_1 \| R_2 \| R_s) + h_{ie}}{1 + h_{fe}}$$
$$= \frac{(27\,\text{K}\|27\,\text{K}\|600) + 1\,\text{K}}{1 + 85} \Rightarrow 18.3\,\Omega.$$
$$R'_o = R_o \| R_E \| R_L \Rightarrow 18.3\,\text{K}\|5.6\,\text{K}\|47\,\text{K} \Rightarrow 18.23\,\Omega,$$

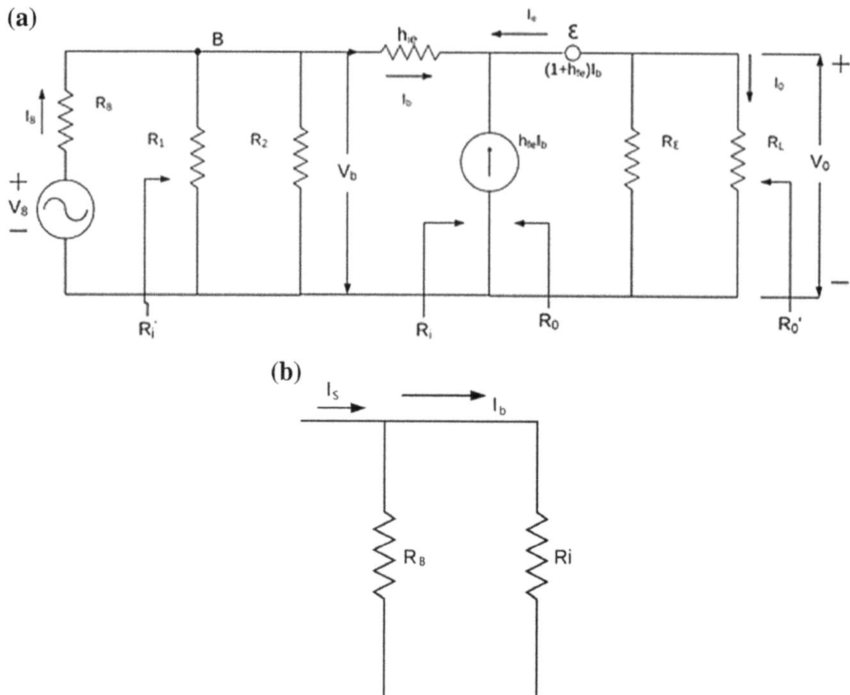

Fig. 3.32 **a** Simplified hybrid model. **b** Circuit using I_b and I_s

(e) **Overall Voltage Gain**:

$$A_{vs} = V_o / V_s = \frac{V_o}{V_b} \times \frac{V_b}{V_s},$$

where $\frac{V_o}{V_b} = A_v$ and $\frac{V_b}{V_s} = \frac{R_i^{'}}{R_i^{'} + R_s}$

$$A_{vs} = \frac{A_v R_i^{'}}{R_i^{'} + R_s} \Rightarrow \frac{0.977 \times 13.09\,\text{K}}{13.09\,\text{K} + 600} \Rightarrow 0.953,$$

(f) **Overall Current Gain**:

$$A_i = \frac{I_o}{I_s} => \frac{I_o}{I_e} \times \frac{I_e}{I_b} \times \frac{I_b}{I_s}$$

where

$$\frac{I_o}{I_e} = \frac{-R_E}{R_E + R_L} \Rightarrow \frac{-5.6\,\text{K}}{5.6\,\text{K} + 47\,\text{K}} \Rightarrow -0.106,$$

$$\frac{I_e}{I_b} = -(1 + h_{fe}) \Rightarrow -86$$

Looking at Fig. 3.32b

$$\frac{I_b}{I_s} = \frac{R_B}{R_B + R_i} \Rightarrow \frac{(27\,\mathrm{K}\|27\,\mathrm{K})}{(27\mathrm{K}\|27\mathrm{K}) + 431.33\mathrm{K}} \Rightarrow 0.03.$$

$$A_{i(\text{for circuit})} = \frac{I_o}{I_s} \Rightarrow (-0.106) \times (-86) \times (0.03) \Rightarrow 0.273.$$

3.10.3 *Analysis of Common Base Circuit Using Simplified Model*

The approximate CB model can be drawn by giving input to emitter, taking output from collector and making base common. Figure 3.33 shows the approximate CB model.

Current Gain (A_i): It is defined as the ratio of output to input currents

$$A_i = \frac{I_o}{I_e} = \frac{-I_c}{I_e} \Rightarrow \frac{-h_{fe} I_b}{-(1 + h_{fe}) I_b}, \quad \because I_c = h_{fe} I_b, \quad \text{and } I_e = -(1 + h_{fe}) I_b$$

$$A_i = \frac{h_{fe}}{1 + h_{fe}} \tag{3.85}$$

The above equation of CB shows that its current gain is always less than one.

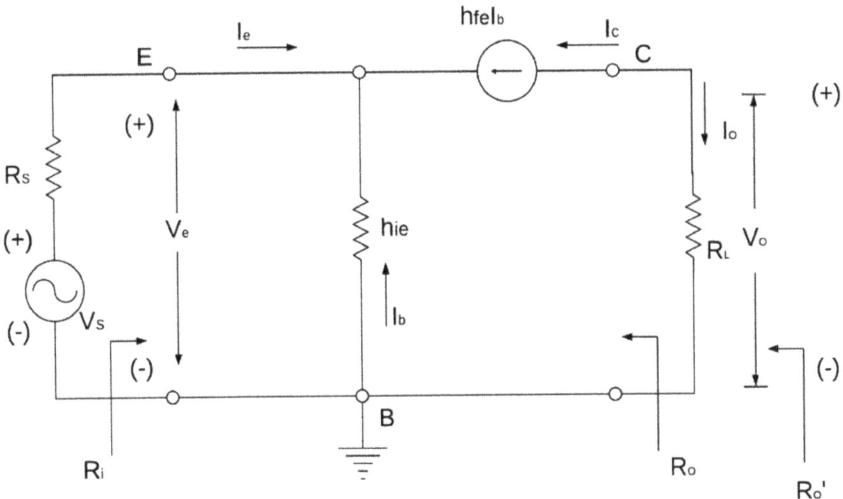

Fig. 3.33 Simplified common base model

Input Resistance (R_i): It is defined as the ratio of the input voltage to input current.

$$R_i = \frac{V_e}{I_e} \Rightarrow \frac{-h_{ie} I_b}{-(1+h_{fe}) I_b} \because V_e = h_{ie} I_b, \text{ and } I_e = -(1+h_{fe}) I_b$$
$$= \frac{h_{ie}}{1+h_{fe}} \tag{3.86}$$

The above equation of CB shows that its input resistance is very low as compared to CE and CC configurations.

Voltage Gain (A_v): It is defined as the ratio of output voltage to input voltages.

$$A_v = \frac{V_o}{V_e} \Rightarrow \frac{I_o R_L}{I_e R_i} \Rightarrow \frac{A_i R_L}{R_i}$$

Substituting value of A_i and R_i, we get

$$A_v = \frac{\frac{h_{fe}}{1+h_{fe}} \times R_L}{\frac{h_{ie}}{1+h_{fe}}} \Rightarrow \frac{h_{fe} R_L}{h_{ie}} \tag{3.87}$$

Output Resistance (R_o): It is the ratio of output voltage to output current at

$$V_s = 0$$
$$R_o = \frac{V_o}{I_c}$$

When $V_s = 0$, the current through the input loop $I_b = 0$, hence $I_c = 0$, and $R_o = \infty$. The output resistance $R_o^{'}$ of the stage taking the load into account is given as

$$R_o^{'} = R_o \| R_L \Rightarrow \infty \| R_L \Rightarrow R_L \tag{3.88}$$

Example 3.6: A common base amplifier, as shown in Fig. 3.34, has the following components: $R_s = 600\,\Omega$, $R_c = 5.6\,\text{K}$, $R_E = 5.6\,\text{K}$, $R_L = 39\,\text{K}$. The transistor parameters are $h_{ie} = 1\,\text{K}$, $h_{fe} = 85$, and $h_{oe} = 2\,\mu\text{A/V}$. Calculate R_i, R_o, A_v, A_{vs}.
Solution: Since $h_{oe} \times (R_C \| R_L) \Rightarrow 2 \times 10^{-6} \times (5.6\,\text{K} \| 39\,\text{K}\Omega) \Rightarrow 9.79 \times 10^{-3}$, which is less than 0.1, we use approximate analysis method. Figure 3.35 shows the simplified hybrid model for the given circuit.

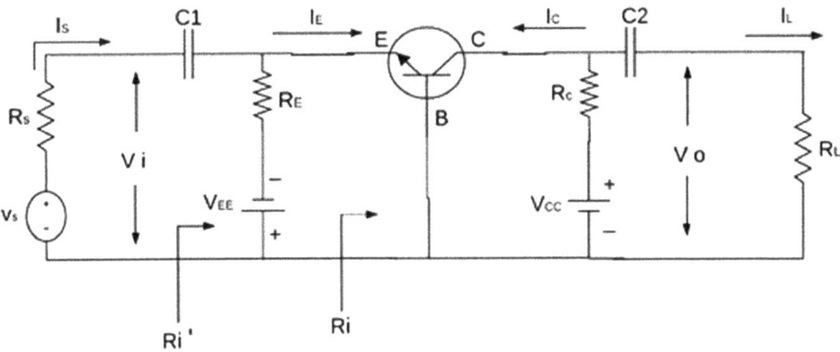

Fig. 3.34 Common base amplifier

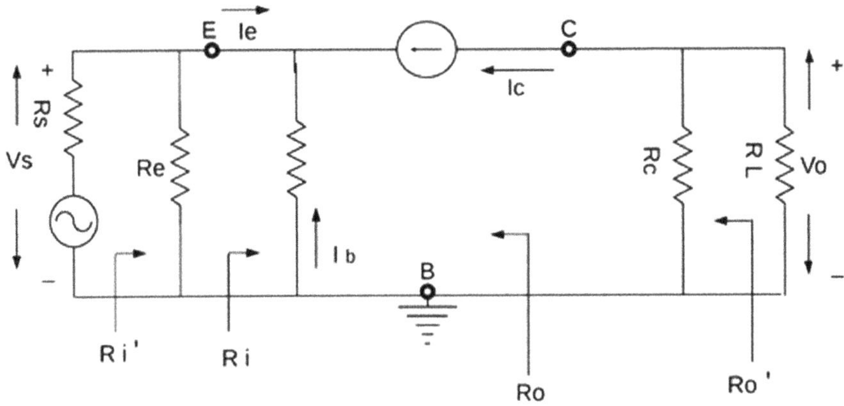

Fig. 3.35 Simplified hybrid model

Current Gain: $(A_i) = \frac{h_{fe}}{1+h_{fe}} \Rightarrow \frac{85}{1+85} => 0.988.$
Input Resistance:

$$(R_i) = \frac{h_{ie}}{1+h_{fe}} \Rightarrow \frac{1000}{1+85} \Rightarrow 11.627$$

$$R_i^{'} = R_i \| R_E \Rightarrow 11.627 \| 5.6\,\text{K} \Rightarrow 11.6\,\Omega$$

Voltage Gain: $(A_v) = \frac{h_{fe}R_L}{h_{ie}} \Rightarrow \frac{85\times(5.6\,\text{K}\|39\,\text{K})}{1\,\text{K}} \Rightarrow 416.23$
Output Resistance: $(R_o) = \infty$

$$R_o^{'} = R_o \| R_L^{'} \Rightarrow R_o \| R_C \| R_L \Rightarrow \infty \| 5.6\,\text{K} \| 39\,\text{K} \Rightarrow 4.89\,\text{K}$$

$$A_{vs} = \frac{V_o}{V_s} \Rightarrow \frac{V_o}{V_e} \times \frac{V_e}{V_s} \Rightarrow A_v \times \frac{V_e}{V_s}$$

where $\frac{V_e}{V_s} = \frac{R_i'}{R_i'+R_s}$

$$A_{vs} = A_v \times \frac{R_i'}{R_i' + R_s} \Rightarrow 416.23 \times \frac{11.6}{11.6 + 600} \Rightarrow 7.89.$$

Summary

- **This chapter included introduction of bipolar transistor, as its application like transistor works as an amplifier.**
- **Ebers–Moll model, computed emitter voltage, and collector voltage.**
- **Describe Miller's theorem.**
- **Analyze low- and high-frequency response of common emitter and common source amplifier configuration.**
- **Hybrid parameters, discuss transistor as a two-port network.**
- **Compute the value of h_{11}, h_{12}, h_{21}, and h_{22}.**
- **Show the h-parameter representation for three types of configuration.**
- **General hybrid equivalent circuit.**
- **Analysis of common emitter circuit using simplified hybrid model.**
- **Analysis of common collector circuit using simplified hybrid model.**
- **Analysis of common base circuit using simplified hybrid model.**

Numerical Problems

Question 1: Explain how transistor works as an amplifier.
Question 2: Discuss low-frequency response of common emitter amplifier configuration.
Question 3: Determine the high-frequency response of the amplifier circuit shown in Fig. 3.36.
Question 4: What are the hybrid parameters.
Question 5: Draw and discuss hybrid model of common base amplifier configuration.
Question 6: Discuss the high-frequency response of common source amplifier configuration.
Question 7: Determine the low-frequency response of the amplifier circuit as shown in Fig. 3.37.

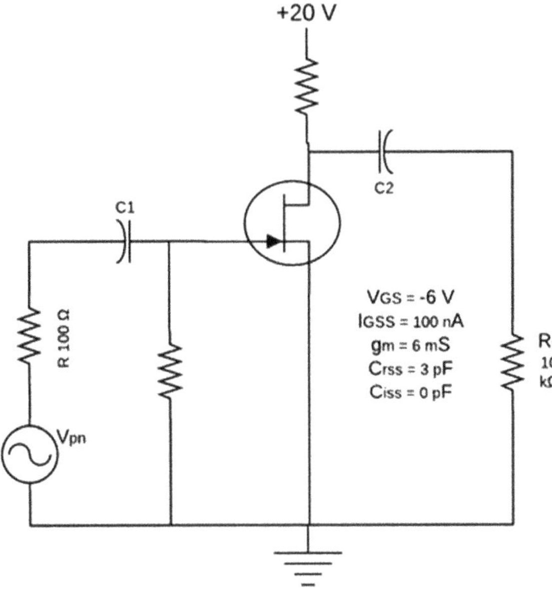

Fig. 3.36 RC network

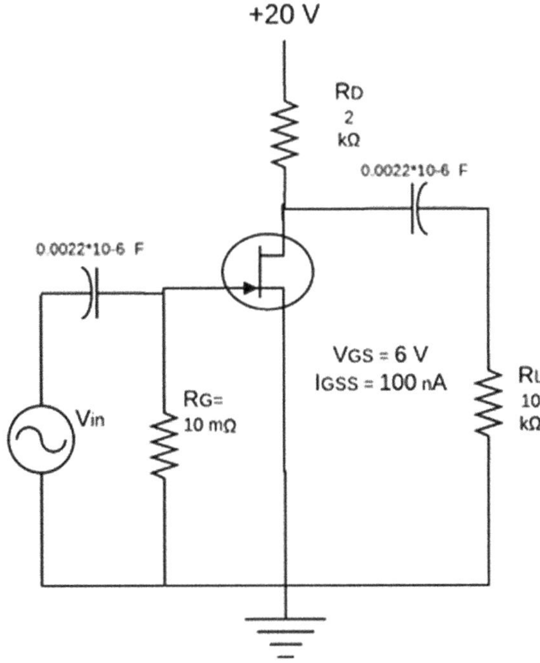

Fig. 3.37 RC network

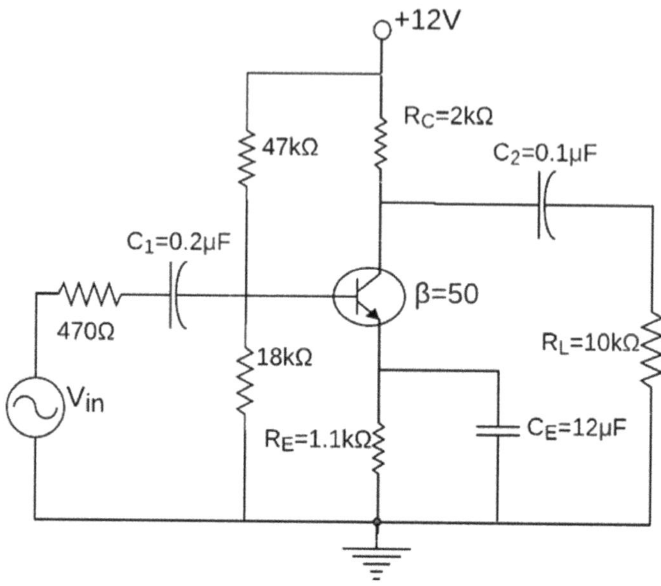

Fig. 3.38 RC network

Question 8: Explain Miller's theorem.
Question 9: Draw and explain Ebers–Moll model.
Question 10: For the circuit shown in Fig. 3.38, determine the low-frequency response.

Chapter 4
Feedback

Learning Objectives

- **Introduction to feedback and their types.**
- **Discuss positive and negative feedback with their properties like transfer gain, stability, etc.**
- **Discuss feedback topologies like series-shunt, series-series, shunt-shunt, and shunt-series.**
- **Analyze voltage shunt, voltage series, current shunt, and current series feedbacks with mathematical expression.**

4.1 Introduction

Feedback makes a significant role in all electronic circuits. It is almost regularly used in the amplifier circuit to optimize its working performance and enhance its quality to make it more ideal. In feedback technique, we use two input signals: one is simply an incoming input signal and other is a part of the output which is feedback to the input terminal. Both these signals may be in phase or out of phase. When incoming input signal and a part of feeded output signal are in phase, the feedback is known as positive feedback. On the other side, when they both signals are in out of phase, the feedback is known as negative feedback. The application of positive feedback is oscillations, while amplifier is the example of negative feedback system.

In this chapter, we discuss the feedback introduction, concept of feedback and show feedback topologies, and show how to modify the characteristics of an amplifier by combining a part of the output signal with the incoming input signal.

G. S. Tomar and A. Bagwari, *Fundamentals of Electronic Devices and Circuits*, Algorithms for Intelligent Systems, https://doi.org/10.1007/978-981-15-0267-5_4

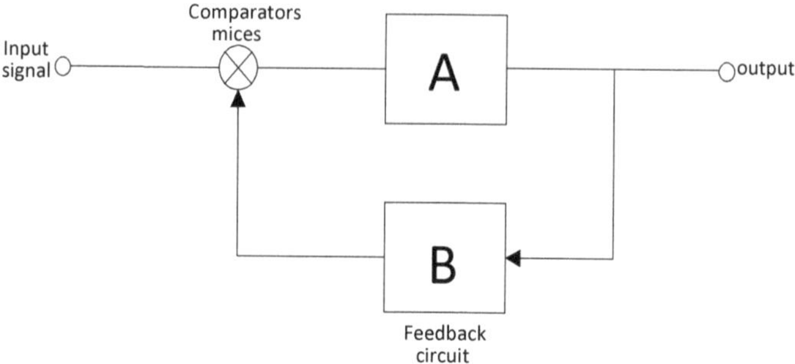

Fig. 4.1 Feedback or closed-loop structure

4.2 Concept of Feedback

Feedback is a mechanism or process in which output signal is looped back to control a system within itself. It may be defined as the process of injecting some energy from the output to the input. In system containing an input and output feeding back, part of the output in such a way to increase the input is positive feedback (Regeneration).

Feeding back part of the output in such a way as to oppose the input is negative feedback (Degeneration). A feedback amplifier basically consists of two parts namely an amplifier circuit and feedback circuit/network. The function of feedback circuit is to return the output energy (may be voltage "*V*" or current "*I*") to the input signal of the circuit.

Figure 4.1 shows the feedback structure, where *A* is forward gain and β is feedback gain or feedback factor. At input side, incoming input signal and loop-backed output signal are mixing with each other with the help of comparator mixer. This comparator mixer can be positive or negative depending on the value of feedback factor (β). This comparator will be either summer or separator, so there are two cases given as

✓ Feedback is negative, if ($\beta < 0$).
✓ Feedback is positive, if ($\beta > 0$).

4.3 Application of Feedback

There are several applications of feedback system as mention below

1. Control theory.
2. Electronic Engineering.
3. Mechanical Engineering.
4. Climate science.
5. Software engineering and computing system.

4.4 Classification of Feedback

There are two types of feedback.

4.4.1. Positive feedback (Oscillator).
4.4.2. Negative feedback (Amplifiers).

4.4.1 *Positive Feedback*

➢ A positive or in-phase feedback signal, where a positive-going wave on the input leads to a positive-going change on the output will amplify the input signal. This is known as positive feedback.
➢ If the feedback signal (i.e., *V* or *I*) is applied in such a way that it is in phase with the input signal and thus increases it, then it is called a positive feedback.
➢ Figure 4.2 shows the structure of positive feedback system, where comparator is working as a summer circuit and adding both the incoming and loop-back signals.

4.4.1.1 Concept of Positive Feedback

This subheading depicts the logic behind developing positive feedback terminology and its advantages.

➢ It is also known as regenerative feedback or direct feedback positive feedback increases noise and distortion.
➢ However, because of its capability of increasing the power of the original signal, it is used in oscillator circuits.

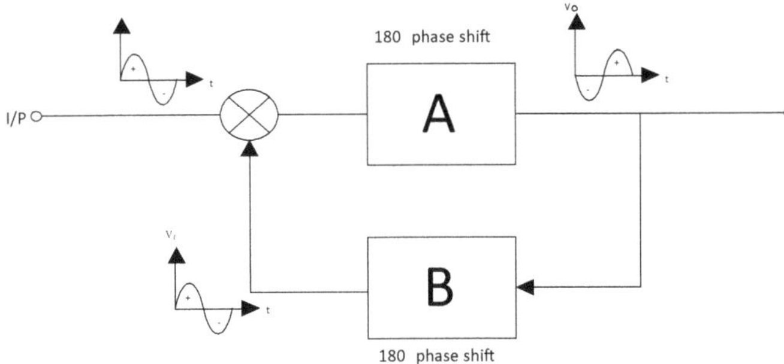

Fig. 4.2 Positive feedback structure

4.4.2 *Negative Feedback*

➢ A negative or out of-phase feedback signal, where a positive-going wave on the input leads to a negative-going change on the output will reduce the amplitude level of the input signal. This is known as negative feedback.
➢ Figure 4.3 shows the structure of negative feedback system, where comparator is working as a separator circuit and subtracting both the incoming and loop-back signals.

4.4.2.1 Concept of Negative Feedback

➢ If the feedback signal (i.e., *V* or *I*) is applied in such a way that it is out of phase with the input signal and thus decreases its input signal, then it is called negative feedback.
➢ It is also called as degenerative feedback or inverse feedback.
➢ It has the following advantages.
Gain stability.
Reduction in non-linear distortion.
Reduction in noise.
Increases bandwidth (that is why most widely used in amplitude).

4.4.2.2 Properties of Negative Feedback

There are four basic properties of negative feedback system given as.

4.4.2.2.1 Transfer Gain.

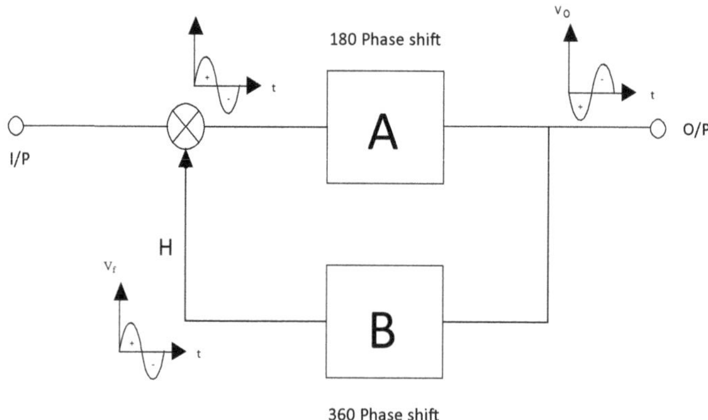

Fig. 4.3 Negative feedback structure

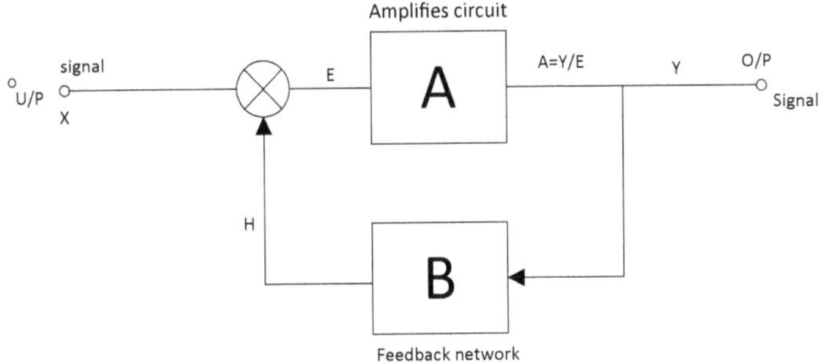

Fig. 4.4 Feedback structure

4.4.2.2.2 Stability in Gain.
4.4.2.2.3 Reduce Noise.
4.4.2.2.4 Reduced Distortion or Interference.

4.4.2.2.1 Transfer Gain

Transfer gain shows the ratio between output value and input value. Transfer function is denoted by A_f. There is a mathematical expression to find out gain of any kind of feedback system is given as,

$$A_f = \frac{\text{Output}}{\text{Input}} = \frac{y}{x} \tag{4.1}$$

$X \to$ Source signal (may be V or I) $Y \to$ Output Signal (V or I)
$E \to$ Input signal (may be V or I) $H \to$ feedback signal (V or I)

Case I: Let assuming the comparator has positive feedback topology in Fig. 4.4, now the transfer gain will be

$$A = \frac{Y}{E} \tag{4.2}$$

$$Y = AE \tag{4.3}$$

$$\text{As, } E = X + H \quad \text{(considering positive feedback)} \tag{4.4}$$

Using Eqs. (4.3) and (4.4), we get

$$Y = A(X + H) \tag{4.5}$$

$$\text{where, feedback factor is } \beta = \frac{H}{Y} \tag{4.6}$$

$$H = \beta Y \tag{4.7}$$

Using Eqs. (4.5) and (4.7), we get

$$Y = A(X + \beta Y) \tag{4.8}$$

$$Y(1 - A\beta) = AX \tag{4.9}$$

$$\frac{Y}{X} = \frac{A}{1 - A\beta} \tag{4.10}$$

Using Eqs. (4.1) and (4.10), we can say that

$$A_{\mathrm{f}} = \frac{Y}{X} = \frac{A}{1 - A\beta} \tag{4.11}$$

$$A_{\mathrm{f}} = \frac{A}{1 - A\beta} \tag{4.12}$$

Case II: Let assuming the comparator has negative feedback topology in Fig. 4.4, now the transfer gain will be

$$A = \frac{Y}{E} \tag{4.13}$$

$$\text{As, } E = X - H \quad \text{(considering negative feedback)} \tag{4.14}$$

Using Eqs. (4.13) and (4.14), we get

$$Y = A(X - H) \tag{4.15}$$

Using Eqs. (4.7) and (4.15), we get

$$Y = A(X - \beta Y) \tag{4.16}$$

$$Y = AX - A\beta Y \tag{4.17}$$

$$Y(1 + A\beta) = AX \tag{4.18}$$

$$\frac{Y}{X} = \frac{A}{1 + A\beta} \tag{4.19}$$

Similarly, using Eqs. (4.1) and (4.19), we can say that

$$A_{\rm f} = \frac{Y}{X} = \frac{A}{1 + A\beta} \tag{4.20}$$

$$A_{\rm f} = \frac{A}{1 + A\beta} \tag{4.21}$$

Where $A_{\rm f}$ is the gain of system with feedback (also called closed-loop gain), A is the open-loop (forward/Amplitude) gain of system without feedback, and β is feedback gain or feedback ratio.

From Eqs. (4.12) and (4.21), we can say that value of $A_{\rm f}$ is depending upon $(1 \mp A\beta)$. If $(1 - A\beta)$ increases, the value of $A_{\rm f}$ decreases and vice versa. $A\beta$ is feedback factor and $(1 \mp A\beta)$ is loop gain.

4.4.2.2.2 Stability in Gain

Stability of gain shows the feedback system stability. It should be as much as possible. This is one of the key properties of negative feedback system makes an important role.

➢ The transfer function gain of amplifier is not constant as depends on the factors such as operating point, temperature.
➢ This lack of stability in amplifier can be reduced by introducing negative feedback (important advantage of negative feedback).

As we know that

$$A_{\rm f} = \frac{A}{1 + A\beta}$$

Differentiating Eq. (4.21) with respect to "*A*," we get

$$\frac{{\rm d}A_{\rm f}}{{\rm d}A} = \frac{\rm d}{{\rm d}A}\left(\frac{A}{1 + A\beta}\right) \tag{4.22}$$

$$\frac{{\rm d}A_{\rm f}}{{\rm d}A} = \frac{(1 + A\beta)\frac{{\rm d}A}{{\rm d}A} - A\frac{\rm d}{{\rm d}A}(1 + A\beta)}{(1 + A\beta)^2} \tag{4.23}$$

$$\frac{{\rm d}A_{\rm f}}{{\rm d}A} = \frac{(1 + A\beta) - A(0 + \beta)}{(1 + A\beta)^2} \tag{4.24}$$

$$\frac{{\rm d}A_{\rm f}}{{\rm d}A} = \frac{1 + A\beta - A\beta}{(1 + A\beta)^2} \tag{4.25}$$

$$\frac{{\rm d}A_{\rm f}}{{\rm d}A} = \frac{1}{(1 + A\beta)^2} \tag{4.26}$$

Further, multiplying and dividing Eq. (4.26) by $A_{\rm f}$, we get

$$\frac{{\rm d}A_{\rm f}}{{\rm d}A} = \frac{1}{(1 + A\beta)^2} \times \frac{A_{\rm f}}{A_{\rm f}} \tag{4.27}$$

$$\frac{dA_f}{A_f} = \frac{1}{(1+A\beta)^2} \times \frac{dA}{A_f} \tag{4.28}$$

Put the value of A_f from Eqs. (4.21) to (4.28), we get

$$\frac{dA_f}{A_f} = \frac{1}{(1+A\beta)^2} \times \frac{dA}{\frac{A}{(1+A\beta)}} \tag{4.29}$$

$$\frac{dA_f}{A_f} = \frac{1}{1+A\beta} \times \frac{dA}{A} \tag{4.30}$$

Where $\frac{dA_f}{A_f}$ is fractional change in amplification with feedback.
$\frac{dA}{A}$ is factional change in amplification without feedback.

- The fractional change in amplification with feedback divided by fractional change in amplifier without feedback is known as sensitivity.

$$\frac{\frac{dA_f}{A_f}}{\frac{dA}{A}} = \text{Sensitivity} = \frac{1}{(1+A\beta)} \tag{4.31}$$

The reciprocal of sensitivity is called de-sensitivity which is

$$(1+A\beta) = \frac{1}{\text{sensitivity}} = \text{Desensitivity} \tag{4.32}$$

Thus, stability increases as de-sensitivity $(1 + A\beta)$ increases.

4.4.2.2.3 Reduce Noise

There are always noise voltages in amplifier which is reduced by a factor $(1+A\beta)$ when negative feedback is applied.

4.4.2.2.4 Reduced Distortion or Interference

To reduce distortion applies negative feedback a factor, $(1 + A\beta)$ is responsible.

4.4.2.3 Drawbacks of Negative Feedback

1. Reduction in input resistance, i.e., "R_{in}" in case of voltage shunt and current shunt feedback amplifier.
2. Increases the output resistance, i.e., "R_{out}" in case of current shunt and current series feedback amplifiers.
3. It reduces gain.
4. More numbers of amplifiers stages are required to obtain the required gain.

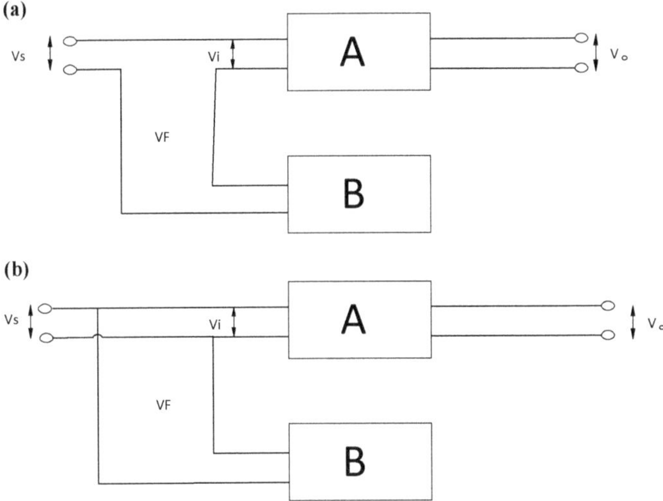

Fig. 4.5 **a** Series connection, **b** Shunt connection

4.5 Basic Feedback Connections

There are mainly two connection exists in feedback systems.

4.5.1. Series connection.
4.5.2. Shunt connection.

4.5.1 Series Connection

In this connection, terminals are connected to each other in series manner. It shows the current flow from one terminal to another will same but the voltage will different. Figure 4.5a shows the series connection, where loop-backed signal and input signal are connected in series to each other.

4.5.2 Shunt Connection

In this connection, terminals are connected to each other in shunt (parallel) manner. It shows the voltage across one terminal and another terminal remains same but there is change in flow of current. Figure 4.5b shows the shunt connection, where loop-backed signal and input signal are connected in parallel to each other.

- **Basic Concept**

This concept will provide a direction and helpful to readers to make desired connections. As we know that in any electronics system, we use either voltage or current as an input supply.

- If voltage exists at input terminal or voltage as an input supply, then input terminal will be connected with feedback terminal in series manner.
- If voltage exists in output side, then output terminal will be connected with feedback terminal in shunt manner.
- If current exists at input terminal or current as an input supply, then input terminal will be connected with feedback terminal in shunt manner.
- If current exists in output side, then output terminal will be connected with feedback terminal in series manner.

4.6 Basic Feedback Topologies

Feedback voltage (V_f) can be proportional to either load voltage or load current. Due to this, V_f can be connected either in series or in shunt with the signal voltage V_s. There are basically four types of feedback systems.

4.6.1 Voltage Series Feedback.
4.6.2 Voltage Shunt Feedback.
4.6.3 Current Series Feedback.
4.6.4 Current Shunt Feedback.

4.6.1 Voltage Series Feedback

Figure 4.6 shows the diagram of voltage series feedback. Using basic concept, we can design it easily. In "voltage series feedback," the notable thing is that the "voltage" will exist in output side and "series" will exist in input side; due to this and considering **basic concept table**, the supply should be voltage because voltage exists in series manner at input terminal as an input supply. So the input-side connection between input terminal and feedback terminal is series manner with voltage supply, while voltage exists output side and using Table 4.1, if voltage exists in output side, the connection between output terminal and feedback terminal will be in shunt manner. Therefore, input and output side will be in series and parallel manner, respectively. This is the easiest method to design any kind of feedback topology. Further, this circuit is called voltage amplifier because forward gain equals to output voltage divided by input voltage.

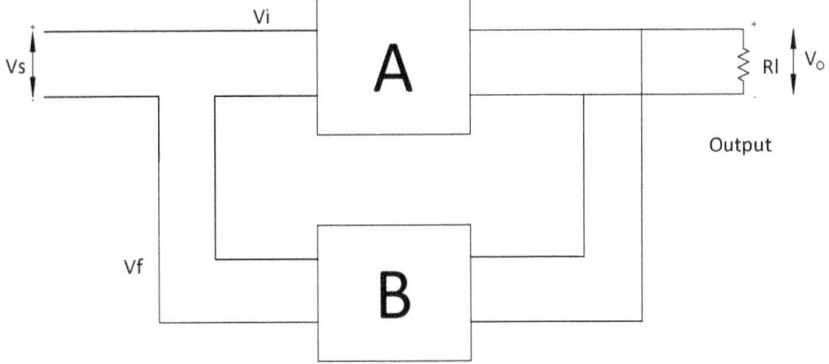

Fig. 4.6 Series shunt feedback topology or Voltage amplifier

Table 4.1 Shows basic concept tables

	Input	Output
Voltage (V)	Series	Shunt
Current (I)	Shunt	Series

$$A_v = \frac{\text{Output Voltage}(V_0)}{\text{Input Voltage}(V_i)} \tag{4.33}$$

It means circuit is amplifying voltage from input side to output side.

- **Voltage Amplifier/Voltage series feedback/Series–shunt feedback topology**

Using Fig. 4.6, we can design its respective amplifier as shown in Fig. 4.7. Voltage is as an input supply connected in series with feedback terminal. So the connection between source voltage (V_S) and input voltage (V_i) is in series manner with each other. R_S and R_i are the internal source resistance and input resistance, respectively. Similarly, at output side, output terminal and feedback terminal are in shunt manner. So, the gain voltage ($A_v V_i$) will be connected in parallel manner with output voltage. We have to calculate the output voltage (V_O) and voltage always taken across any load

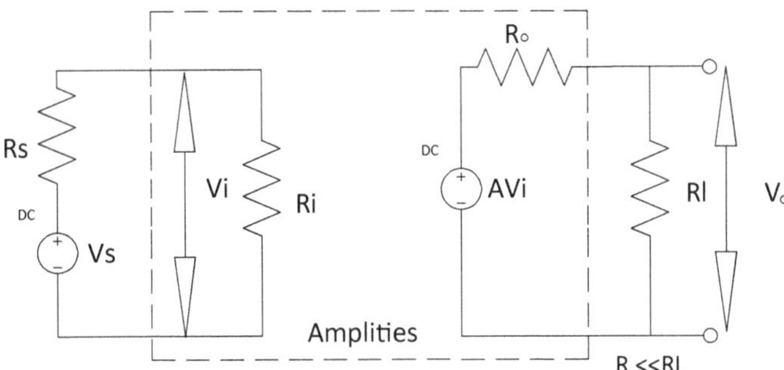

Fig. 4.7 Thevenin's equivalent circuit of an amplifier

so the load resistance (R_L) will be in shunt manner. Now, the Thevenin's equivalent circuit of a two-port between which represents an amplifier

If the amplifier input resistance $R_i \gg R_s$ shows R_i is open circuit and R_S is short circuit, where R_S is supply source resistance then we have

$$V_s \cong V_i. \tag{4.34}$$

If external load $R_L \gg R_o$(output resistance) of the amplifier, R_L is open circuit and R_O is short circuit then,

$$A_v V_i = V_o \tag{4.35}$$

$$A_V = \frac{V_O}{V_i} \tag{4.36}$$

Using Eqs. (4.34) and (4.36), we get

$$A_V = \frac{V_O}{V_S} \tag{4.37}$$

Where

A_v is voltage gain.

So, we can say that an amplifier produces an output voltage (V_O) proportional to source voltage (V_S) and it does not depend upon source load R_s such circuits are called voltage amplifier circuit.

Note: An ideal voltage amplifier input resistance $\boldsymbol{R_i = \infty}$***, output resistance*** $\boldsymbol{R_O = 0}$***, while practically*** $\boldsymbol{R_i}$ ***should be large and*** $\boldsymbol{R_O}$ ***should be small.***

4.6.2 Voltage Shunt Feedback

Figure 4.8 shows the diagram of voltage shunt feedback. Using basic concept, we can design it easily. In "voltage shunt feedback," the notable thing is that the "voltage" will exist in output side and "shunt" will exist in input side; due to this and considering **basic concept table,** the supply should be current. According to Table 4.1, current exists in shunt manner at input terminal as an input supply. So the input-side connection between input terminal and feedback terminal is shunt manner with current supply, while voltage exists output side and using Table 4.1, if voltage exists in output side, the connection between output terminal and feedback terminal will be in shunt manner. Therefore, input and output sides will be in series manner respectively.

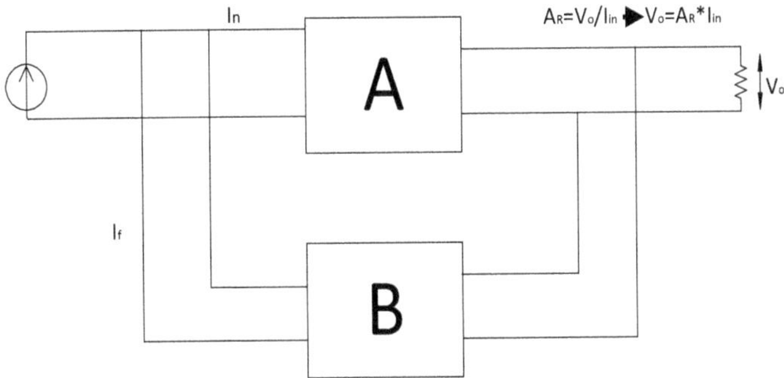

Fig. 4.8 Shunt–shunt feedback topology or Trans-resistance amplifier

- **Trans-resistance amplifier/Voltage shunt feedback/Shunt-shunt feedback topology**

Using Fig. 4.8, we can design its respective amplifier as shown in Fig. 4.9. Current is as an input supply, connected in parallel with feedback terminal. So the connection between source current (I_S) and input current (I_i) is in parallel manner with each other. R_S and R_i are the internal source resistance and input resistance, respectively. Similarly, at output side, output terminal and feedback terminal are in shunt manner. So the gain voltage ($A_R I_{in}$) will be connected in parallel manner with output voltage. We have to calculate the output voltage (V_O) and voltage always taken across any load so the load resistance (R_L) will be in shunt manner. Now, Norton's equivalent circuit of a two-port between which represents an amplifier.

Trans-resistance amplifier is one which provides output voltage (V_0) proportional to supply current (I_s) without depending on R_s and R_L. Norton's equivalent circuit of Trans-resistance amplifier.

For practical Trans-resistance amplifier, we must have $R_s \gg R_{in}$ and $R_0 \gg R_L$

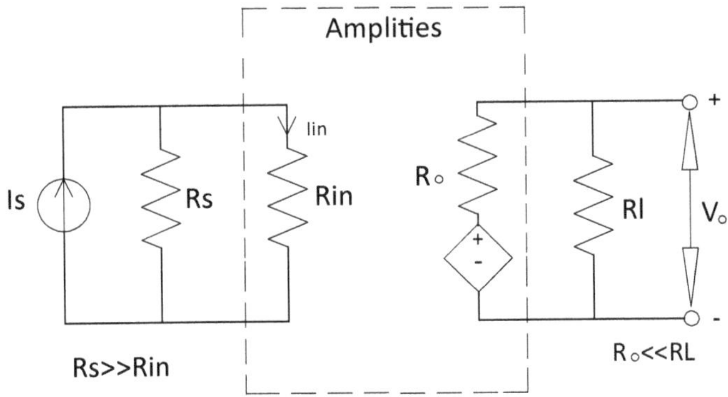

Fig. 4.9 Norton's equivalent circuit of an amplifier

$$\text{If } R_s \gg R_{in},\ I_s = I_{in} \text{ (where } R_s \text{ in O.C.)} \tag{4.38}$$

$$\text{If } R_0 \gg R_L,\ V_0 = A_R I_{in} \text{ (Where } R_0 \text{ is S.C.)} \tag{4.39}$$

Using Eqs. (4.38) and (4.39), we get

$$V_0 = A_R I_s \tag{4.40}$$

$$A_R = \frac{V_0}{I_s} \tag{4.41}$$

Equation (4.41) means output voltage is depending upon supply current or I_s. The common emitters circuit may be considered as Trans-resistance amplifier if $R_L \gg R_0$.

4.6.3 *Current Shunt Feedback*

Figure 4.10 shows the diagram of current shunt feedback. Using basic concept, we can design it easily. In "current shunt feedback," the notable thing is that the "current" will exist in output side and "shunt" will exist in input side; due to this and considering **basic concept table,** the supply should be current. According to Table 4.1, current exists in shunt manner at input terminal as an input supply. So the input-side connection between input terminal and feedback terminal is shunt manner with current supply, while current exists output side and using Table 4.1, if current exists in output side, the connection between output terminal and feedback terminal will be in series manner. Therefore, input and output sides will be in shunt and series manner respectively.

- **Current amplifier/Current shunt feedback/Shunt-series feedback topology**

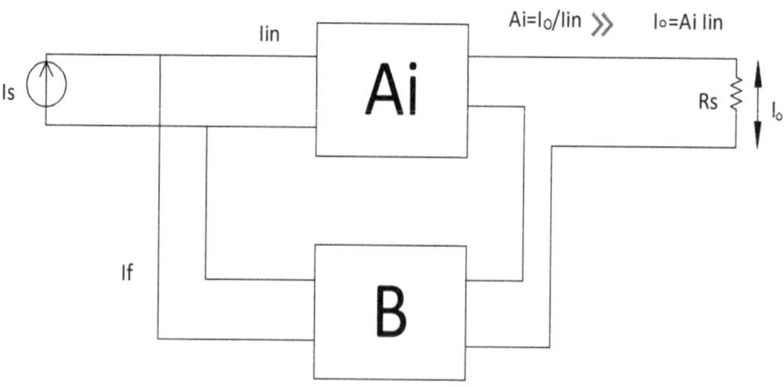

Fig. 4.10 Shunt-series feedback topology or Current amplifier

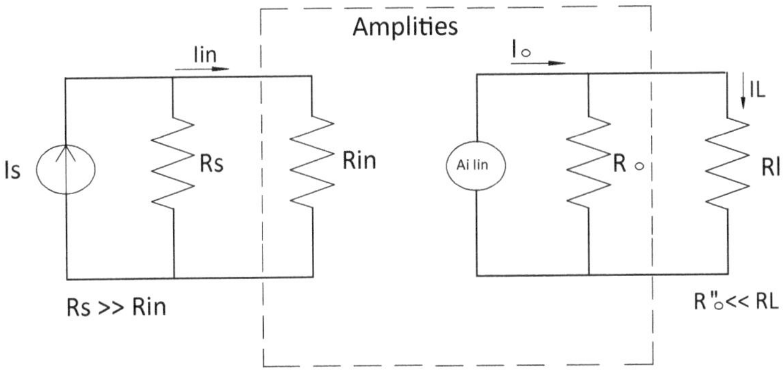

Fig. 4.11 Norton's equivalent of a current amplifier

Using Fig. 4.10, we can design its respective amplifier as shown in Fig. 4.11. Current is as an input supply, connected in parallel with feedback terminal. So the connection between source current (I_S) and input current (I_i) is in parallel manner with each other. R_S and R_i are the internal source resistance and input resistance, respectively. Similarly, at output side, output terminal and feedback terminal are in series manner. So the gain current ($A_i I_{in}$) will be connected in series manner with output current. We have to calculate the output current (I_O) and current always taken in series so the load resistance (R_L) will be in series manner. Now, Norton's equivalent circuit of a two-port between which represents an amplifier

The ideal current amplifier produces an output current (I_0) proportional to the input current (I_{in}) and proportional factor does not depends on R_s or R_L.

For practical current amplifier, we must have $R_s \gg R_{in}$,

$$I_s = I_{in} \text{ (as } R_s \text{ is open circuit)} \tag{4.42}$$

If $R_0 \gg R_L$,

$$I_0 = A_i I_{in} \text{ (as } R_0 \text{ is O.C.)} \tag{4.43}$$

Using Eqs. (4.42) and (4.43), we get

$$I_0 = A_i I_s \tag{4.44}$$

$$A_i = \frac{I_0}{I_s} \tag{4.45}$$

Equation (4.45) shows output current (I_o) proportional to input current.

Note: An ideal current amplifier must have $R_{in} = 0$ and $R_0 = \infty$ practically, the input resistance should be as small as possible and the output resistance must be as large as possible.

4.6.4 *Current Series Feedback*

Figure 4.12a shows the diagram of current series feedback. Using basic concept, we can design it easily. In "current series feedback," the notable thing is that the "current" will exist in output side and "series" will exist in input side; due to this and considering **basic concept table,** the supply should be voltage. According to Table 4.1, voltage exists in series manner at input terminal as an input supply. So the input-side connection between input terminal and feedback terminal is series manner with voltage supply, while current exists output side and using Table 4.1, if current exists in output side, the connection between output terminal and feedback terminal will be in series manner. Therefore, input and output sides will be in series and series manners, respectively.

- **Series-series feedback or trans-conductance amplifier**

Using Fig. 4.12a, we can design its respective amplifier as shown in Fig. 4.12b. Voltage is as an input supply, connected in series with feedback terminal. So the connection between source voltage (V_S) and input voltage (V_i) is in series manner with each other. R_S and R_i are the internal source resistance and input resistance, respectively. Similarly, at output side, output terminal and feedback terminal are in

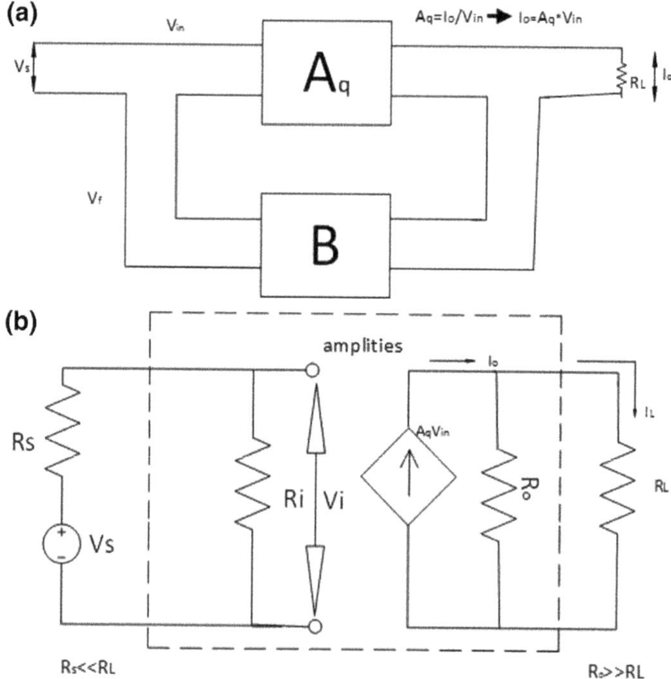

Fig. 4.12 **a** Series-series feedback topology or Trans-conductance amplifier, **b** Thevenin's equivalent of Trans-conductance amplifier

series manner. So the gain current ($A_G V_{in}$) will be connected in series manner with output current. We have to calculate the output current (I_O) and current always taken in series so the load resistance (R_L) will be in series manner. Now, the Thevenin's equivalent circuit of a two-port between which represents an amplifier.

An ideal Trans-conductance amplifier produces an output current (I_0) which is proportional to input voltage (V_{in}) and independent of the magnitude of R_s and R_L.

- **Thevenin's equivalent of trans-conductance amplifier**

Ideally, this amplifier must have an infinite input resistance ($R_{in} = \infty$) and infinite output resistance ($R_0 = \infty$), i.e., $R_{in} = R_0 = \infty$ but practically, both these values should be as high as possible

$$\text{If } R_s \gg R_i,\ V_s = V_{in} (\text{as } R_s \text{ is open circuit}) \quad (4.46)$$

$$\text{If } R_0 \gg R_L,\ I_L = A_G V_{in}\ (\text{as } R_0 \text{is open circuit}) \quad (4.47)$$

Using Eqs. (4.46) and (4.47), we get

$$I_L = A_G V_s \quad (4.48)$$

$$A_G = \frac{I_L}{V_s} \quad (4.49)$$

Equation (4.49) shows load current (I_L) depends on supply voltage (V_S).

Question 4.1: An amplifier with voltage gain of 60 dB uses 1/20 of its output in negative feedback. Determine the gain with feedback in dB.

Solution: $A = 60\text{dB}$

$$20 \log_{10} A = 60$$

$$\log_{10} A = 60/20$$

$$\log_{10} A = 3$$

Now, $$A = 10^3$$

$$A = 1000.$$

$$\beta = \frac{1}{20}$$

$$A_f = \frac{A}{1 + A\beta} = \frac{1000}{1 + 1000 \times \frac{1}{20}}$$

And, $$A_f = \frac{1000}{51}$$

$$A_f = 19.60.$$

$$A_f = 20 \log_{10} 19.60$$

$$A_f = 25.84\,\text{dB}$$

Question 4.2: The voltage gain of an amplifier without feedback is 60 dB; it reduces to 40 dB with feedback. Determine feedback factor.
Solution: $A = 60$ dB and $A_\text{f} = 40\,\text{dB}$.

$$20\log_{10} A = 60 \quad \text{and} \quad 20\log_{10} A_\text{f} = 40$$
$$\log_{10} A = 3\log_{10} A_\text{f} = 2$$
$$A = 10^3 \quad A_\text{f} = 10^2$$
$$A = 1000. \quad A_\text{f} = 100.$$

As we know,
$$A_\text{f} = \frac{A}{1 + A\beta}$$
$$100 = \frac{100}{1 + 1000\beta}$$
$$100 + 100000\,\beta = 1000$$
$$\beta = \frac{900}{100000}$$
$$\beta = 0.009.$$

Now,
$$\beta = 20\log_{10} 0.009$$
$$\beta = -40.91\,\text{dB}.$$

Question 4.3: A single stage transistor amplifier has voltage gain without feedback 600 and 50 with feedback. Find out output voltage of output which is feedback to input side.
Solution: $A = 600$ and $A_\text{f} = 50$

$$A_\text{f} = \frac{A}{1+A\beta}$$
$$50 = \frac{600}{1+600\beta}$$
$$50 + 3000\beta = 600$$

Now, $\beta = \frac{550}{3000}$
$$\beta = 0.01833$$
$$\beta \text{ in percentage} = 0.01833 \times 100$$
$$= 1.833\%.$$

Question 4.4: An amplifier with a negative feedback provides an output voltage of 5 V with an input of 0.2 V, removing feedback it requires only 0.1 V input supply to provide the same output. Calculate:

1. Gain with feedback.
2. Gain without feedback.
3. Feedback ratio.

Solution: $V_\text{O} = 5\,\text{V}$, and $V_\text{i} = 0.2\,\text{V}$

(i) Gain with feedback "A_f"

$$A_f = \frac{V_0}{V_i}$$

$$A_f = \frac{5.0}{0.2}$$

$$A_f = 25.$$

(ii) Gain without feedback "A"

$$A = \frac{\text{output}}{\text{input}}$$

$$A = \frac{50}{0.1}$$

$$A = 50.$$

(iii) Feedback ratio "β"

$$A_f = \frac{A}{1 + A\beta}$$

$$25 = \frac{50}{1 + 50\beta}$$

$$25 + 1250\beta = 50$$

$$\beta = \frac{25}{1250}$$

$$\beta = 0.02.$$

Question 4.5: Negative feedback of $\beta = 0.002$ is applied to an amplifier of gain 1000. Calculate the change in overall gain of the feedback amplifier. If the internal amplifier is subjected to a gain reduce to 15%.

Solution: $\beta = 0.002$, and $A = 1000$

$$A_f = \frac{A}{1 + A\beta} = \frac{1000}{1 + 1000 \times 0.002}$$

$$A_f = \frac{1000}{3}$$

$$A_f = 333.33.$$

$$A_{new} = (A - A \times 15\%)$$

$$A_{new} = A - A \times \frac{15}{100}$$

$$A_{new} = 1000 - 1000 \times \frac{15}{100}$$

$$A_{new} = 1000 - 150$$

$$A_{new} = 850.$$

$$A_{\text{fnew}} = \frac{A_{\text{new}}}{1+\beta A_{\text{new}}} \Rightarrow \frac{850}{1+0.002\times 850}$$

Now, $A_{f\text{new}} = \frac{850}{2.5}$

$$A_{\text{fnew}} = 314.81.$$

Percentage change in overall gain of feedback

$$A_{\text{f}}^{'} = \frac{A_{\text{f}} - A_{\text{fnew}}}{A_{\text{f}}} = \frac{333.33-314.81}{333.33}$$

amplifier $A_{\text{f}}^{'} = \frac{18.51}{333.33}$

$$A_{\text{f}}^{'} = 0.0555$$

$$A_{\text{f}}^{'} = 5.55\%.$$

4.7 Determination of Loop Gain

Loop gain shows the gain within the closed loop, used to analyze and measure the feedback system performance. Figure 4.13 depicts the feedback amplifier configuration.

➢ It is denoted by "T" which is the product of $A\beta$, where A is the forward gain and β is feedback factor or feedback gain.
➢ It is an important parameter in the stability of feedback circuit.

- **Approach or procedure to finding the loop gain**

➢ In Fig. 4.14, the loop has been broken at the input of the amplifier and a test signal is applied at this input.
➢ Now the amplifier output is given by,

$$S_0 = AS_{\text{t}} \tag{4.50}$$

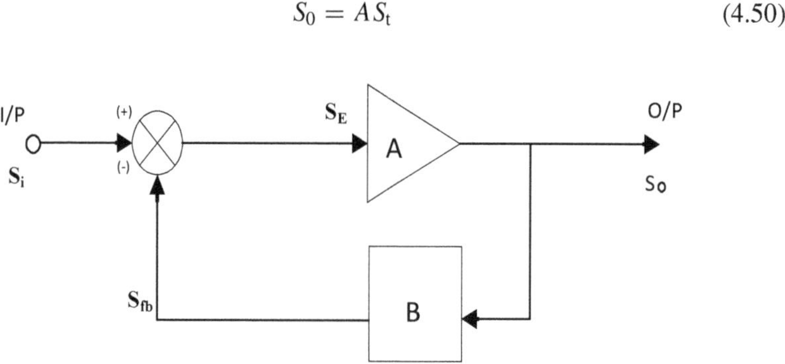

Fig. 4.13 Ideal feedback amplifier configuration

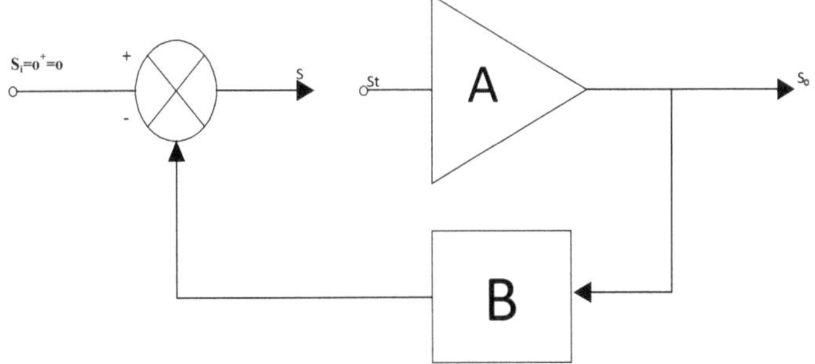

Fig. 4.14 Same configuration with the loop broken after amplifier input

$$\text{And, } \left(A = \frac{S_0}{S_t}\right) \tag{4.51}$$

Now the feedback signal is given by,

$$\text{As, } \beta = \frac{S_{fb}}{S_0} \tag{4.52}$$

$$S_{fb} = \beta S_0 \tag{4.53}$$

Using Eqs. (4.50) and (4.53), we get

$$S_{fb} = \beta A S_t \tag{4.54}$$

- The signal S_r is called return signal, earlier it is said error signal S_E.
- Now, we have

$$S_r = S_i - S_{fb} \tag{4.55}$$

Let $S_i = 0^+ = 0$(if there is no signal)
So we can say that

$$S_r = -S_{fb} \tag{4.56}$$

Where, negative symbol indicates that the type of feedback is negative feedback.

- Now from Eqs. (4.54) and (4.56), we can say that

$$S_r = -A\beta S_t \tag{4.57}$$

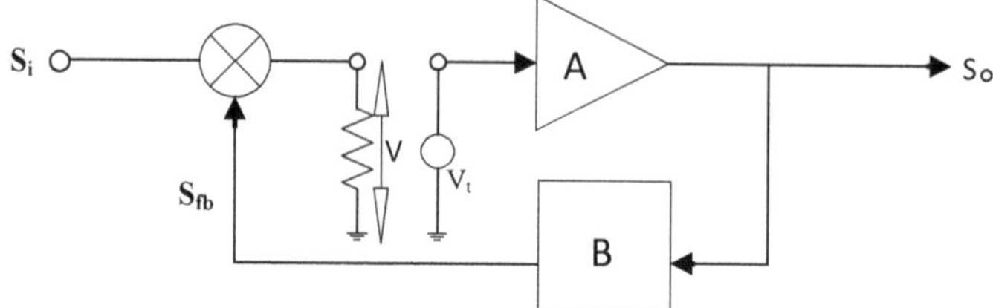

Fig. 4.15 Same configuration with the loop broken in terms of impedance

$$\frac{S_r}{S_t} = -A\beta \tag{4.58}$$

$$A\beta = -\frac{S_r}{S_t} = T \tag{4.59}$$

➢ This shows that the ratio of return signal S_r and test signal S_t is equal to negative of the loop gain factor $(A\beta)$
➢ When we break the loop, the conditions exiting before breaking the loop should remain unchanged.

- **Now, in terms of impedance**

Consider Fig. 4.15, after applying the test voltage 'V_t', the return voltage 'V_r' is measured and the loop gain (T) is applied as under

$$A\beta = T = -\frac{V_r}{V_t} \tag{4.60}$$

Case I: If resistance (R) is constant, then loop gain will be

$$A\beta = T = -\frac{I_r}{I_t} \tag{4.61}$$

Case II: If current (I) is constant, then loop gain will be

$$A\beta = T = -\frac{R_r}{R_t}. \tag{4.62}$$

4.8 Stability Problem

➢ Using negative feedback, our system becomes stable and works as an amplifier while applying negative feedback to positive feedback, our system becomes unstable and works as an oscillator.
➢ We know that for a basic feedback configuration, the ideal closed-loop transfer function (gain) is given by

$$A_f = \frac{A}{1 + A\beta} \tag{4.63}$$

➢ The open-loop gain "A" is not constant. It is a function of individual transistor parameter (L and C), so it is a function of frequency.
➢ Hence, the closed-loop gain A_f also becomes a function of frequency and mathematical expression is

$$A_f(s) = \frac{A(s)}{1 + \beta A(s)} \tag{4.64}$$

Where $\beta A(s)$ is loop gain denoted by $= T(s)$.
And, s is frequency denoted by $= j\omega$.
➢ Now

$$A_f(s) = \frac{A(s)}{1 + T(s)} \tag{4.65}$$

Hence, the loop gain $T(s)$ may be represented as $T(s) = T(j\omega)$ and expressed as under

$$A_f(j\omega) = \frac{A(j\omega)}{1 + T(j\omega)} \tag{4.66}$$

$$T(j\omega) = |T(j\omega)| \times \angle\phi \tag{4.67}$$

Because

$$a + jb = \sqrt{a^2 + b^2} \times \tan^{-1}\left(\frac{b}{a}\right) \tag{4.68}$$

$$= |a + jb| \times \angle\phi \tag{4.69}$$

➢ **Observations**

1. For insuring stability, the value of closed-loop gain which is $A_f(j\omega)$ at a given frequency should be small enough so that the Barkhausen criteria should not be satisfied and the oscillation should not begin.

2. The stability of feedback circuit is a function of loop gain $T(j\omega)$ if $|T(j\omega)| = 1$ and $\angle\phi = 180°$, then $T(j\omega) = -1$ and according to equation, the value of $A_{\mathrm{f}}(j\omega)$ becomes infinite and the system loss its stability as shown in Fig. 4.16.
3. If $|T(j\omega)| > 1$ and $\angle\phi = 180°$, then $T(j\omega) =$ negative and system becomes unstable as shown in Fig. 4.17
4. If $|T(j\omega)| < 1$ and $\angle\phi = 180°$, then $T(j\omega) =$ positive and system becomes stable as shown in Fig. 4.18.

Questions 4.6: An amplifier gain changes by ±10% using negative feedback, the amplifier is to be modified to provide a gain of 100 with ±0.1% variation obtain the required open-loop gain of amplifier and the amount of negative feedback.

Solution: $A_{\mathrm{f}} = 100,\ \frac{\mathrm{d}A_{\mathrm{f}}}{A_{\mathrm{f}}} = \pm 0.1\%,$ and $\frac{\mathrm{d}A}{A} = \pm 10\%.$

Now,

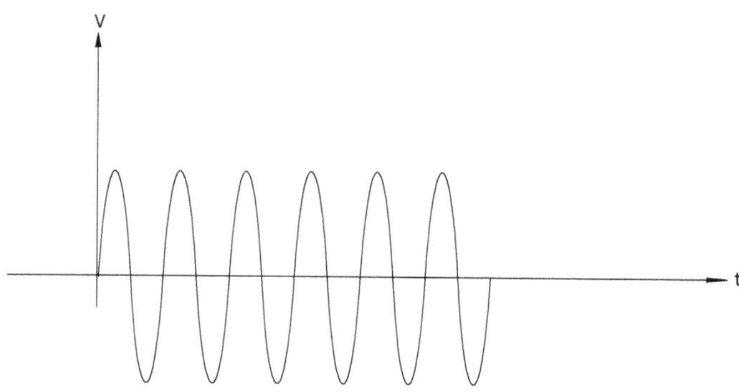

Fig. 4.16 Sustained oscillations

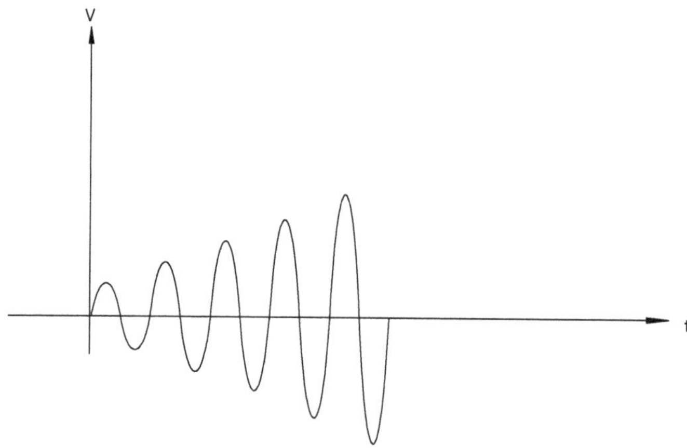

Fig. 4.17 Growing oscillations

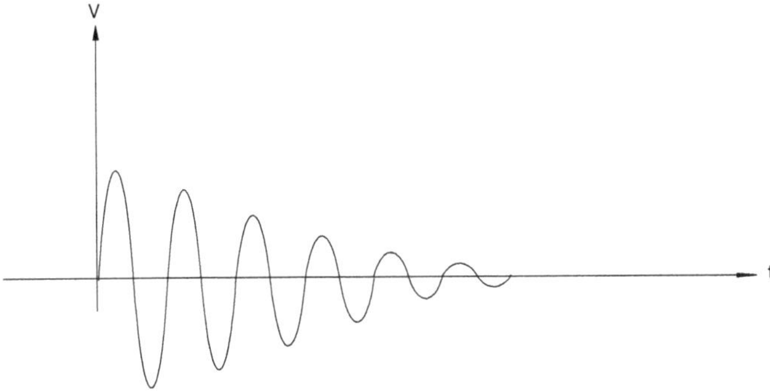

Fig. 4.18 Damped or die out oscillation

$$\frac{dA_f}{A_f} = \frac{1}{1 + A\beta} \times \frac{dA}{A}$$
$$0.1\% = \frac{1}{1 + A\beta} \times 10\%$$
$$\frac{0.1}{1000} = \frac{1}{1 + A\beta} \times \frac{10}{100}$$
$$1 + A\beta = 100$$
$$A\beta = 99.$$

(i)
$$A_f = \frac{A}{1 + A\beta}$$
$$100 = \frac{A}{1 + 99}$$
$$100 = \frac{A}{100}$$
$$A = 10^4.$$

(ii)
$$\text{Amount of negative feedback} = \frac{A_f}{A} = \frac{100}{10^4} = 10^{-2} = 0.01.$$

Question 4.7: An open-loop gain $A = 400$ $\boldsymbol{\beta} = 0.1$ open-loop gain change $\frac{dA}{A} = 20\%$, compute $\frac{dA_f}{A_f}$.

Solution: $A = 400$, $\beta = 0.1$, and $\frac{dA}{A} = 20\%$.

$$\frac{dA_f}{A_f} = \frac{1}{1 + A\beta} \times \frac{dA}{A}$$

$$\frac{dA_f}{A_f} = \frac{1}{1 + (400) \times (0.1)} \times \frac{20}{100}$$

$$\frac{dA_f}{A_f} = \frac{20}{41} \times \frac{1}{100}$$

$$\frac{dA_f}{A_f} = \frac{0.4878}{100}$$

$$\frac{dA_f}{A_f}_{\text{in percentage}} = 0.0048 \times 100$$

$$\frac{dA_f}{A_f} = 0.48\%.$$

Summary

- **This chapter included introduction of feedback.**
- **Positive feedback and negative feedback.**
- **Discuss various properties such as transfer gain and stability, etc.**
- **Relationship between forward gain A, closed-loop gain A_f, and feedback factor β.**
- **Discuss basic feedback connection, series and shunt.**
- **Basic feedback topologies like voltage series, voltage shunt, current series, and current shunt.**
- **Discuss voltage amplifier, current amplifier, Trans-conductance, Trans-resistance amplifiers.**
- **Calculate loop gain.**
- **Discuss stability problem in feedback system.**

Numerical Problems

Question 1: What is feedback shown the relationship between A_f, A and β where A_f is gain with feedback, A is forward gain, and β is feedback ratio.
Question 2: An amplifier with negative feedback providers; an output voltage of 5 V with an input voltage of 0.2 V; on removing feedback, it requires only 0.1 V input to provide the same output, calculate

1. Gain without feedback.
2. Gain with feedback.
3. Feedback ratio.

Question 3: The negative feedback of $\beta = 0.002$ us applied to an amplifier of gain 1000; calculate the change in overall gain of the feedback amplifier if the internal amplifier is subjected to a gain reduce to 15%.
Question 4: Explain shunt-series feedback and show the current amplifier circuit also.
Question 5: Explain stability in gain, how stability depending on de-sensitivity. Show the mathematical relationship.
Question 7: What is feedback and explain positive feedback and negative feedback.
Question 8: What it voltage amplifier and draw its equivalent circuit?
Question 9: Explain current series feedback. Draw circuit diagram?
Question 10: What is loop gain?
Questions 11: Amplifier gain ±10% using negative feedback and $A = 1000, \frac{dA_f}{A_f} = \pm.1\%$, Compute $\frac{dA}{A} = ?$ and $\beta = ?$

Chapter 5
Oscillators

Learning Objectives

- **Introduction to oscillator, mathematical expression, and applications.**
- **Derive and discuss Barkhausen criterion for oscillation.**
- **Discuss sinusoidal and non-sinusoidal oscillators, and their classifications.**
- **Introduce and derive frequency response of various oscillators such as Wien-bridge, RC Phase shift, Lumped oscillator, Hartley oscillator, Colpitts oscillator, Clapp oscillator, and Crystal oscillator.**
- **Discuss quality of factor with mathematical expression.**

5.1 Introduction

Oscillator is one of the useful and important devices which is used in electronic and electrical engineering systems. Oscillator is used to generate AC voltage without applying an AC input signal. To generate AC voltage, the circuit is supplied input signal from a DC source. There are various applications of oscillator such as in amplitude and frequency modulation scheme. Local oscillators are used to provide high frequency radio frequency carrier signal. The oscillator circuits are also used in the transmitter to produce the radio frequency (RF) carrier signal. Other applications are used in the form of clocks in digital systems such as microcomputers and in the sweep circuits used in TV sets and oscilloscopes. The frequency range of oscillator varies from few Hz to several GHz.

Generally, it is said that an oscillator creates a sinusoidal signal, though, it does not generate energy but simply works as an energy converter. It basically converts unidirectional current drawn from a DC source of supply into AC of preferred frequency. Sometimes, it is known as an inverter; because, oscillator is just doing reverse work of a rectifier. In this chapter, we will discuss about oscillator introduction, concepts

G. S. Tomar and A. Bagwari, *Fundamentals of Electronic Devices and Circuits*, Algorithms for Intelligent Systems, https://doi.org/10.1007/978-981-15-0267-5_5

and conditions of oscillation, Barkhausen criterion, its applications, and classification of oscillator. Finally, we will show the frequency response of each and every oscillator.

5.2 Concept of Oscillators

- Any circuit which is used to generate alternate current (AC) voltage without AC input is called oscillator.
- In Fig. 5.1a, at input side instead of AC voltage, we applied direct current (DC) voltage.
- In oscillator, we use positive feedback signal; due to this, it makes or increases distortion and instability. It covers the frequency range of server Hz to GHz.

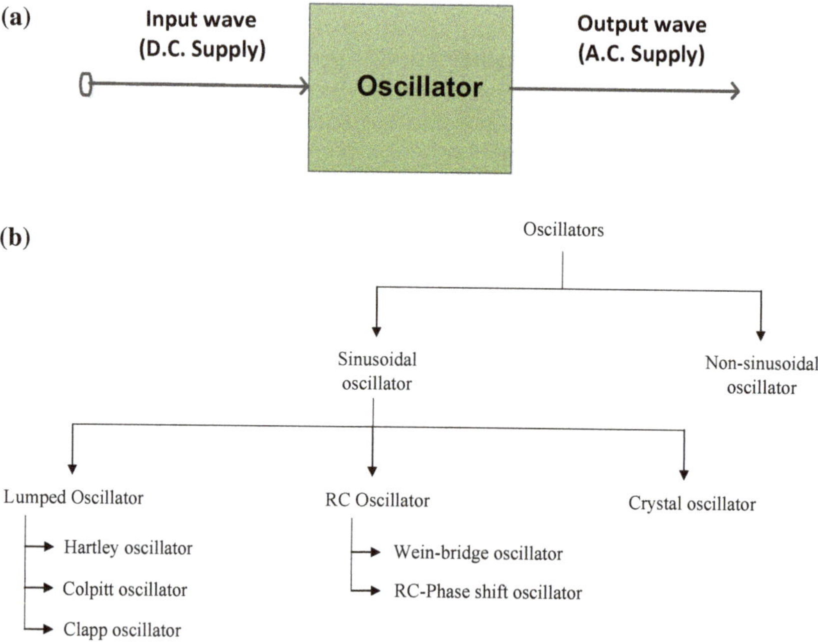

Fig. 5.1 **a** Oscillator structure. **b** Classifications of oscillator

5.3 Barkhausen Criterion

- An oscillator is basically an amplifier which does not have any AC input, but it operates on the principle of positive feedback to generate an AC signal at its output.
- Thus, it is clear that an amplifier can work as an oscillator if positive feedback is made to exit, however, positive feedback not always guarantee oscillator.
- An amplifier will work as an oscillator if and only if it satisfies a set of conditions called Barkhausen criterion.
- Barkhausen criterion states that the frequency of sinusoidal oscillator determined by the condition that the loop phase shift is hence $A\beta = 1$(where, $A\beta$ is loop gain) the overall gain of the system with feedback

$$A_{\mathrm{f}} = \frac{1}{1 - A\beta} \quad \text{(for oscillator with positive feedback)}$$

5.3.1 *Conditions of Oscillation*

- According to Barkhausen criteria, the condition for oscillator is that

  1. $A\beta = 1, \angle\phi = 180°$.
  2. $A\beta > 1, \angle\phi = 180°$.

  So, we can say that conditions for oscillation are $A\beta \geq 1$ and $\angle\phi = 180°$.
  Note: If $A\beta < 1$ and $\angle\phi = 180°$, oscillator will work as an amplifier.

5.4 Application of Oscillator

1. Used in microcomputer.
2. Used in oscilloscope.
3. Used in the form of clock in digital system.
4. Used to generate a carrier wave.

5.5 Classification of Oscillator

5.5.1 Sinusoidal or Harmonic Oscillator

This kind of oscillator provides an output having sine wave the output frequency range varies from 20 Hz to 1 GHz.

5.5.2 Non-sinusoidal or Relaxation Oscillator

The output waveform having a square, rectangular, and saw-tooth waveform. Such oscillator can provide output at frequency ranging from 0 Hz to 20 MHz. Figure 5.1b depicts the classification of oscillator.

5.6 Types of Sinusoidal Oscillator

5.6.1. RC Oscillator: This oscillator consists of R and C to generate low-frequency signal, and hence it is also known as audio frequency (AF) oscillator. Such oscillators are Wien-bridge and RC phase shift oscillators.
5.6.2. Lumped oscillator (Tuned circuit oscillator): This oscillator consists of L and C which are used to generate high-frequency signal, and hence it is also known as radio frequency (RF) oscillator. Such oscillators are Hartley, Colpitts, and Clapp oscillators.
5.6.3. Crystal Oscillator: Here, we use Quartz crystal and generate high stability output signal with frequency up to 10 MHz, for example, Pierce Oscillator.
5.6.4. Negative Resistance Oscillator: Oscillator which uses negative resistance characteristic of the device, for example, tunnel diode oscillator.

5.6.1 RC Oscillator

- In RC oscillator, it is generally used for generating audio frequency signal as they provide good frequency stability and waveform.
- Further with the advent of IC technology, RC network is only feasible solution as it is very difficult to make too high value inductance in an IC.
- Two commonly used RC networks are:

  5.6.1.1. Wien-bridge Oscillator.
  5.6.1.2. RC Phase shift Oscillator.

5.6.1.1 Wien-bridge Oscillator

- Generally in an oscillator, amplifier stage introduces 180° phase shift and feedback network introduces again 180° phase shift to obtain a phase of 360° around a loop.
- But Wien-bridge oscillator uses a non-negative amplifier and hence does not provide any phase shift during amplifier stage as total phase shift required is 0° or $2\pi n$ radian.
  where $n = 1, 2, 3, 4, \ldots$
- In Fig. 5.2a, the output of amplifier is applied between the terminal 1 and 3 which is input to the feedback network.
- While the amplifier input is supplied from diagonal terminal 2 and 4, this is output from the feedback network.

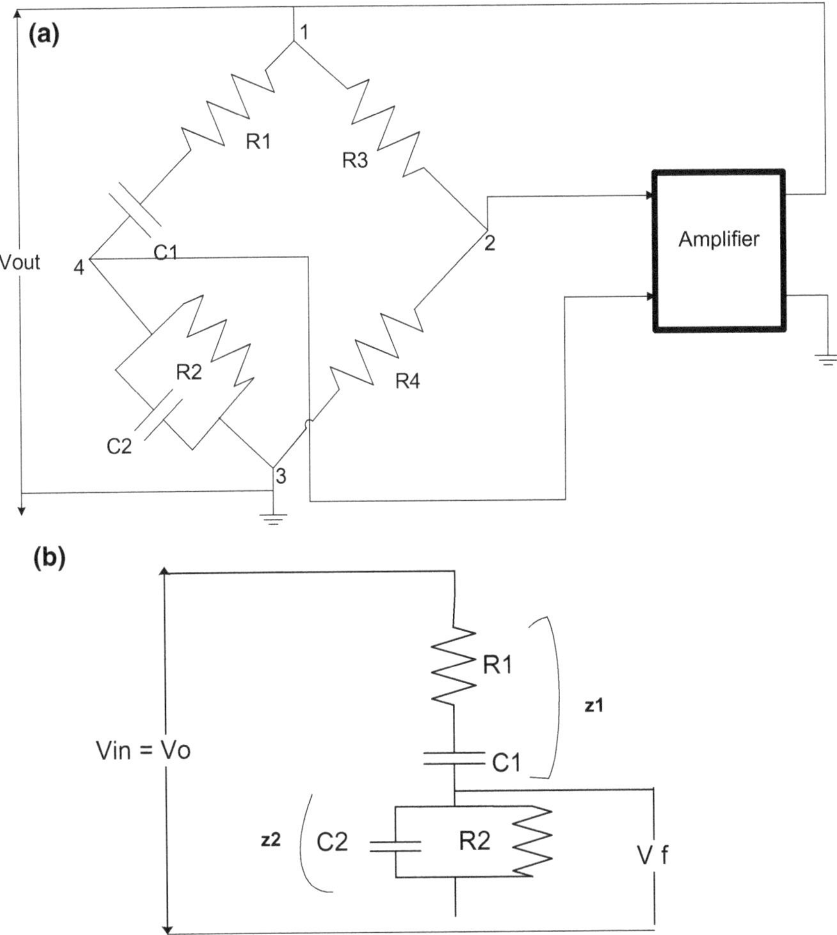

Fig. 5.2 **a** Basic circuit of Wien-Bridge oscillator. **b** Feedback network of Wien-bridge

- Thus, the amplifier get the supplied input signal through the output of the Wein Bridge network as a feedback.
- The two arms of bridge R_1 and C_1 are in series and R_2 and C_2 are in parallel and known as frequency-sensitive arms.

- **Expression or derivation for frequency of oscillation**:

Using Fig. 5.2b, we can calculate the values of Z_1 and Z_2 given as

Case I: R_1 and X_{C1} both are connected in series, then the resultant value will be

$$Z_1 = R_1 + X_{C1} = R_1 + \frac{1}{jwc_1} \tag{5.1}$$

$$Z_1 = \frac{1 + jwR_1C_1}{jwc_1} \tag{5.2}$$

Case II: R_2 and X_{C2} both are connected in parallel, then the resultant value will be

$$Z_2 = R_2 \| X_{C2} \Rightarrow R_2 \| \frac{1}{jwc_2} \tag{5.3}$$

$$Z_2 = \frac{R_2}{1 + jwR_2C_2} \tag{5.4}$$

Now applying KCL in Fig. 5.2b, we get

$$V_{\text{in}} = Z_1 I + Z_2 I \Rightarrow I(Z_1 + Z_2) \tag{5.5}$$

$$I = \frac{V_{\text{in}}}{Z_1 + Z_2} \tag{5.6}$$

We can compute the feedback voltage V_{f} across Z_2, which is

$$V_{\text{f}} = Z_2 \times I \tag{5.7}$$

Using Eqs. (5.6) and (5.7), we can say that

$$V_{\text{f}} = Z_2 \times \left(\frac{V_{\text{in}}}{Z_1 + Z_2} \right) \tag{5.8}$$

$$\frac{V_{\text{f}}}{V_{\text{in}}} = \frac{Z_2}{Z_1 + Z_2} \tag{5.9}$$

Using Fig. 5.2b, we know that V_{in} is equal to V_{o}, then

$$\frac{V_f}{V_{in}} = \frac{V_f}{V_0} = \beta \tag{5.10}$$

and,

$$\beta = \frac{Z_2}{Z_1 + Z_2} \tag{5.11}$$

Put the value of Z_1 from Eq. (5.2) and value of Z_2 from Eqs. (5.4) to (5.11), we get

$$\beta = \frac{\left(\frac{R_2}{1+jwc_2R_2}\right)}{\left(\frac{1+jwR_1C_1}{jwc_1}\right) + \left(\frac{R_2}{1+jwc_2R_2}\right)} \tag{5.12}$$

$$\beta = \frac{\left(\frac{R_2}{1+jwc_2R_2}\right)}{\frac{(1+jwR_1C_1)(1+jwc_2R_2)+R_2jwC_1}{(jwc_1)\times(1+jwR_2C_2)}} \tag{5.13}$$

$$\beta = \frac{R_2jwc_1}{(1 + jwC_1R_1)(1 + jwC_2R_2) + R_2jwC_1} \tag{5.14}$$

$$\beta = \frac{R_2jwC_1}{1 + jwC_2R_2 + jwC_1R_1 + j^2w^2CR_1C_2R_2 + R_2jwC_1} \tag{5.15}$$

$$\beta = \frac{R_2jwC_1}{1 + jwC_1R_1 + jwC_2R_2 - w^2C_1R_1C_2R_2 + R_2jwC_1} \tag{5.16}$$

$$\beta = \frac{jwR_2C_1}{\left(1 - w^2C_1R_1C_2R_2\right) + jw(R_1C_1 + R_2C_2 + R_2C_1)} \tag{5.17}$$

Multiplying and dividing Eq. (5.17) by

$$(a - jb), \text{ where } a = \left(1 - w^2C_1R_1C_2R_2\right), \text{ and } \quad b = (R_1C_1 + R_2c_2 + R_2C_1)w$$

Now,

$$\beta = \frac{jwC_1R_2\left[\left(1 - w^2R_1C_1R_2C_2\right) - jw(R_1C_1 + R_2C_2 + R_2C_1)\right]}{\left[\left(1 - w^2R_1C_1R_2C_2\right) + jw(R_1C_1 + R_2C_2 + R_2C_1)\right] \times \left[\left(1 - w^2R_1C_1C_2\right) - jw(R_1C_1 + R_2C_2 + R_2C_1)\right]} \tag{5.18}$$

$$\beta = \frac{\left(jwC_1R_2\left(1 - w^2R_1C_1R_2C_2\right) - j^2w^2C_1R_2(R_1C_1 + R_2C_2 + R_2C_1)\right)}{\left(1 - w^2R_1C_1R_2C_2\right)^2 - j^2w^2(R_1C_1 + R_2C_2 + R_2C_2)^2} \tag{5.19}$$

$$\beta = \frac{jwC_1R_2\left(1 - w^2R_1C_1R_2C_2\right) + w^2C_1R_2(R_1C_1 + R_2C_2 + R_2C_1)}{\left(1 - w^2R_1C_1R_2C_2\right)^2 + w^2(R_1C_1 + R_2C_2 + R_2C_1)^2} \tag{5.20}$$

Let real part of Eq. (5.20) is $A = w^2C_1R_2(R_1C_1 + R_2C_2 + R_2C_1)$ & and imaginary part of Eq. (5.20) is $jB = jwC_1R_2\left(1 - w^2R_1C_1R_2C_2\right)$.

$$\text{Then, } \beta = \frac{jB + A}{\left(1 - w^2R_1C_1R_2C_2\right)^2 + w^2(R_1C_1 + R_2C_2 + R_2C_1)^2} \tag{5.21}$$

Now, assuming that imaginary part of Eq. (5.21) is zero, i.e., $jB = 0$ or negligible.

$$jwC_1R_2\left(1 - w^2R_1C_1R_2C_2\right) = 0 \tag{5.22}$$

$$1 - w^2R_1C_1R_2C_2 = 0 \tag{5.23}$$

$$w^2R_1C_1R_2C_2 = 1 \tag{5.24}$$

$$w = \sqrt{\frac{1}{R_1R_2C_1C_2}} \tag{5.25}$$

As we know that $w = 2\pi f$, then using Eq. (5.25), we get

$$2\pi f = \sqrt{\frac{1}{R_1R_2C_1C_2}} \tag{5.26}$$

$$f = \frac{1}{2\pi\sqrt{R_1R_2C_1C_2}} \tag{5.27}$$

Equation (5.27) shows the standard frequency response of Wien-bridge oscillator, where R_1, R_2 are the arms resistors and C_1, C_2 are the capacitors.

- Special case, if $R_1 = R_2 = R$ and $C_1 = C_2 = C$, using Eq. (5.27), the oscillator frequency response will be

$$f = \frac{1}{2\pi\sqrt{R.R.C.C}} \tag{5.28}$$

$$f = \frac{1}{2\pi RC} \tag{5.29}$$

Using Eq. (5.20) at $R_1 = R_2 = R$ and $C_1 = C_2 = C$, the gain of feedback network becomes

$$\beta = \frac{jwRC\left(1 - w^2R^2C^2\right) + w^2RC(3RC)}{\left(1 - w^2R^2C^2\right)^2 + w^2(3RC)^2} \tag{5.30}$$

Putting $w = \frac{1}{RC}$ in Eq. (5.30), we get

$$\beta = \frac{j \times \frac{1}{RC} \times RC\left(1 - \frac{R^2C^2}{(RC)^2}\right) + \frac{1}{(RC)^2} RC(3RC)}{\left(1 - \frac{1}{R^2C^2} \times R^2C^2\right) + \frac{1}{R^2C^2}(3RC)^2} \tag{5.31}$$

$$\beta = \frac{0+3}{9} \tag{5.32}$$

$$\beta = \frac{1}{3} \tag{5.33}$$

Conditions for Wien-bridge Oscillation

According to Barkhausen criteria, an oscillator will oscillate if and only if $A\beta \geq 1$, putting the value of β, we get

$$A \times \frac{1}{3} \geq 1 \tag{5.34}$$

$$A \geq 3 \tag{5.35}$$

It implies that for Wien-bridge oscillator circuit, the amplifier gain (A) should be greater than or equal to 3 in order to provide oscillation.

Advantage of Wien-Bridge Oscillator

- By varying the two capacitor values simultaneously, different frequency ranges can be obtained.

Question 5.1 For the output amplifier-based Wien-bridge $R = 5.1\,\text{k}\Omega$, $C = 1\,\text{nF}$, $R_f = 12\,\text{K}\Omega$, $R_1 = 5.1\,\text{K}\Omega$. Determine whether the circuit will oscillate or not. Obtain the output frequency.

Solution

(i) $R_f = R_2 = 12\,\text{K}\Omega$, as we know that the gain formula for non-inverting terminal is

$$A = \frac{V_0}{V_{in}} = \left(1 + \frac{R_f}{R_1}\right)$$

$$A = 1 + \frac{120}{51}$$

$$A = \frac{57}{17}$$

$$A = 3.35.$$

Condition for oscillator in case of Wien-bridge is $A \geq 3$ (in this question value of A is greater than 3) so the circuit will oscillate.

(ii) Now the oscillator frequency will be

$$f = \frac{1}{2\pi RC}$$

$$f = \frac{1}{2\pi\left(5.1 \times 10^{3}\right)\left(1 \times 10^{-9}\right)}$$

$$f = 31206.85\,\text{Hz}$$

$$f = 31.20\,\text{KHz}.$$

Question 5.2 The frequency-sensitive arm of Wien-bridge oscillator uses $C_1 = C_2 = 0.001\,\mu\text{f}$, $R_1 = 10\,\text{K}\Omega$, frequency varies from 10 to 50 KHz. Compute the value of R_2.

Solution The formula for frequency response of Wien-bridge oscillator is

$$f = \frac{1}{2\pi RC}$$

$$f = \frac{1}{2\pi\sqrt{R_1 R_2 C_1 C_2}}$$

$$(2\pi f)^2 = \frac{1}{R_1 R_2 C_1 C_2}$$

$$R_2 = \frac{1}{(2\pi f)^2 R_1 C_1 C_2}$$

(i) When $f = 10$ kHz.

$$R_2 = \frac{1}{(2\pi\, 10)^2 10 \times 10\left(0.001 \times 10^{-6}\right)^2}$$

$$R_2 = 25.33\,\text{K}\Omega.$$

(ii) When $f = 50$ kHz.

$$R_2 = \frac{1}{\left(2\pi \times 50 \times 10^3\right)^2 \times 10 \times 10^3 \times \left(0.001 \times 10^{-6}\right)^2}$$

$$R_2 = 1.013\,\mathrm{K}\Omega.$$

5.6.1.2 RC Phase Shift Oscillator

RC phase shift oscillator basically consists of an amplifier and a feedback network consisting of resistors and capacitors arranged in ladder fashion. Hence, such an oscillator is also known as ladder-type RC phase shift oscillator. Figure 5.3a shows FET-based phase shift oscillator with feedback network.

In RC-based oscillator, we use three RC elements. There is 60° phase shifting between each and every RC elements as shown in Fig. 5.3b.

Figure 5.3c shows the system design of RC phase shift oscillator. In this circuit, FET-based amplifier is used which converts input signal to 180° phase shift output signal. The output loop-backed signal goes to input terminal through three RC networks which provides total 180° phase-shifted output signal to input terminal. Here, each RC network is responsible for providing 60° phase shifting and the net resultant change in the phase is 180° due to the use of three RC networks.

Figure 5.3d is the extended version of Fig. 5.3a, in which we use current dependent source ($g_m \times V_{GS}$) as a supply to feedback network. r_D, R_D, and R' all are internal resistance in the given circuit.

After solving Fig. 5.3a, we get Fig. 5.3e, which is updated form. Here, all internal resistances are replaced by R_0. Where $R_0 = (r_D \mid\mid R_D \mid\mid R')$, and the source supply is voltage source denoted by V_0. The value of V_0 is depending on current dependent source ($g_m \times V_{GS}$) and output resistance (R_0), i.e. $V_0 = g_m \times V_{GS} \times R_0$.

Now, applying Thevenin's Equivalent theorem in Fig. 5.3f, if $R_0 \cong 0$ (means short circuit), then applying KCL inside the circuit

$$V_0 = I_1 \times X_C + (I_1 - I_2) \times R \tag{5.36}$$

$$V_0 = I \times (R + X_C) - R \times I_2 \tag{5.37}$$

$$0 = I_2 \times X_C + (I_2 - I_3) \times R + (I_2 - I_1) \times R \tag{5.38}$$

$$0 = I_2 \times (2R + X_C) - R \times I_3 - R \times I_1 \tag{5.39}$$

$$0 = X_C \times I_3 + R \times I_3 + R \times (I_3 - I_2) \tag{5.40}$$

$$0 = -R \times I_2 + I_3 \times (2R + X_C) \tag{5.41}$$

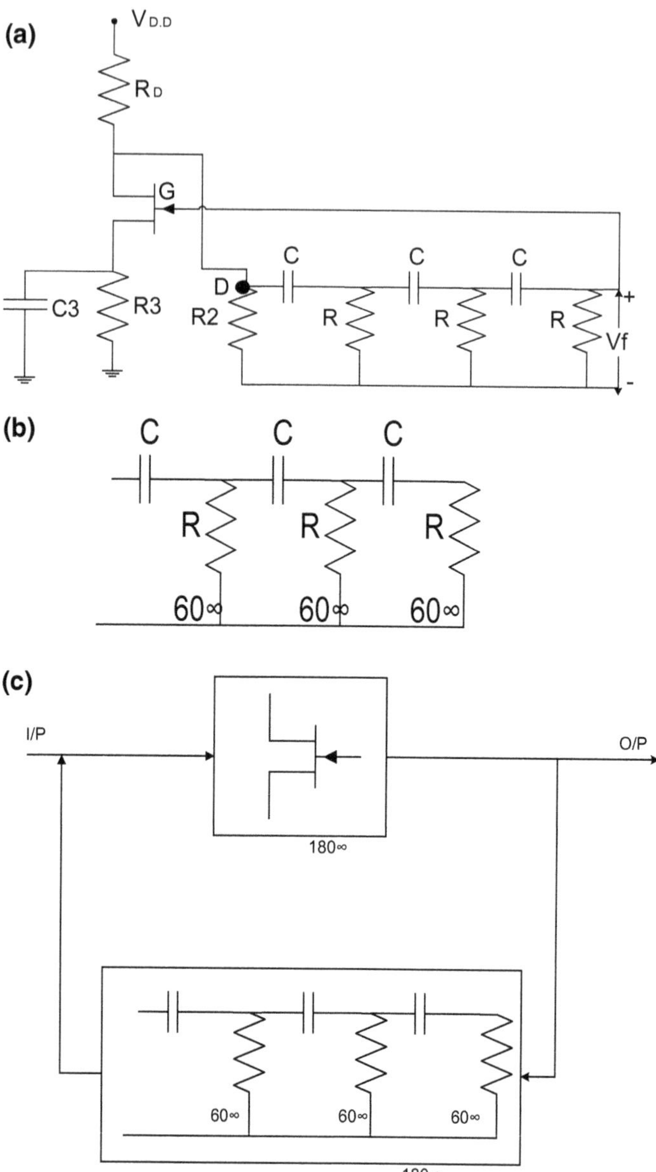

Fig. 5.3 **a** FET-based phase shift oscillator with feedback network. **b** Phase shift network. **c** Architecture of RC phase shift oscillator. **d** Simplified form of Fig. 5.3 (c). **e** Simplified form after combined the internal resistances. **f** Applying KCL

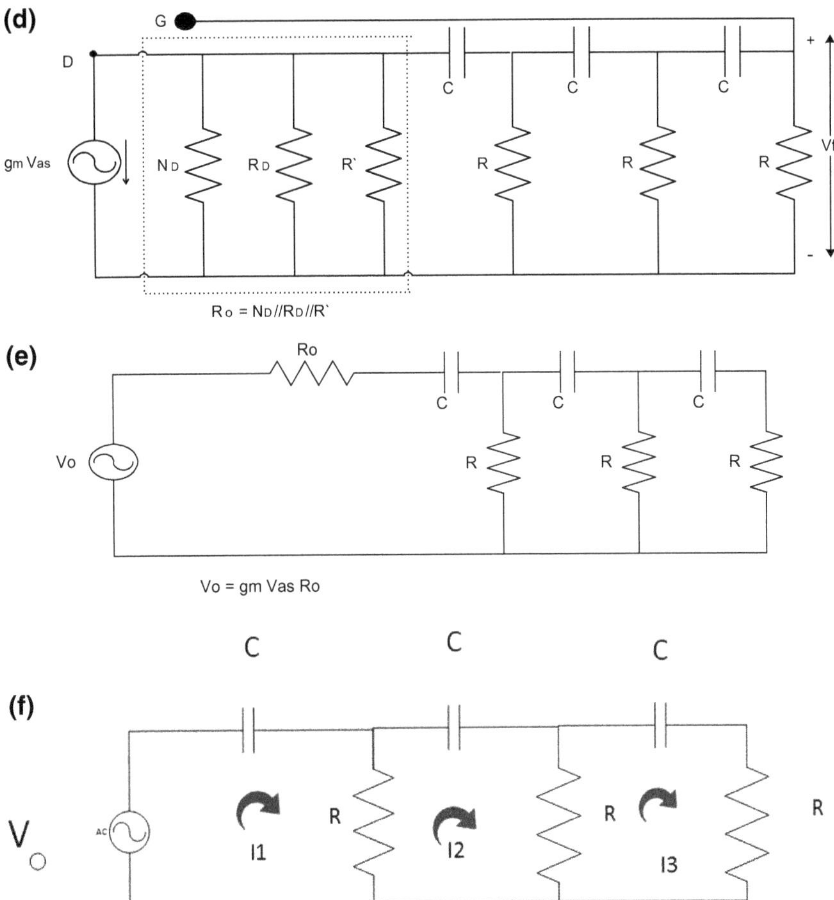

Fig. 5.3 (continued)

Representing Eqs. (5.37), (5.39), and (5.41) in matrix form, we get

$$\begin{bmatrix} (R + X_C) & -R & 0 \\ -R & 2R + X_C & -R \\ 0 & -R & (2R + X_C) \end{bmatrix} \begin{bmatrix} I_1 \\ I_2 \\ I_3 \end{bmatrix} = \begin{bmatrix} V_0 \\ 0 \\ 0 \end{bmatrix} \tag{5.42}$$

Let assuming that $\frac{X_C}{R} = i\alpha$
Now,

$$D = R^3 \begin{bmatrix} 1 + i\alpha & -1 & 0 \\ -1 & 2 + i\alpha & -1 \\ 0 & -1 & 2 + i\alpha \end{bmatrix} \tag{5.43}$$

$$D = R^3[(1 + \alpha i)\{(22 + i\alpha(2 + i\alpha)(2 + i\alpha) + 1)\} - 1[0 - \{-(2 + i\alpha) + 0)\} \tag{5.44}$$

$$D = R^3\left[\left(1 - 5\alpha^2\right) + i\left(6\alpha - \alpha^3\right)\right] \tag{5.45}$$

Therefore, the feedback voltage will be taken across load resistor "R" as

$$V_f = I_3 \times R \tag{5.46}$$

$$\text{And },I_3 = \frac{N_3}{D} \tag{5.47}$$

$$N_3 = R^2\begin{bmatrix} 1 + i\alpha & -1 & V_0 \\ -1 & 2 + i\alpha & 0 \\ 0 & -1 & 0 \end{bmatrix} \tag{5.48}$$

$$N_3 = R^2[0 - 0 + V_0(1 - 0)] \tag{5.49}$$

$$N_3 = R^2 V_0 \tag{5.50}$$

Now putting Eqs. (5.45) and (5.50) in Eq. (5.47), we get

$$I_3 = \frac{V_0 R^2}{R^3\left[\left(1 - 5\alpha^2\right) + i\left(6\alpha - \alpha^3\right)\right]} \tag{5.51}$$

$$I_3 = \frac{V_0}{R\left[\left(1 - 5\alpha^2\right) + i\left(6\alpha - \alpha^3\right)\right]} \tag{5.52}$$

Now using Eq. (5.46), we can say that $V_f = I_3 \times R$, thus I_3 will be

$$I_3 = \frac{V_f}{R} = \frac{V_0}{R\left[\left(1 - 5\alpha^2\right) + i\left(6\alpha - \alpha^3\right)\right]} \tag{5.53}$$

$$\frac{V_f}{V_0} = \frac{1}{\left[\left(1 - 5\alpha^2\right) + i\left(6\alpha - \alpha^3\right)\right]} \tag{5.54}$$

As we know that the feedback factor (β) is ratio of feedback voltage (V_f) to output voltage (V_0), so we can write it as

$$\beta = \frac{V_f}{V_0} = \frac{1}{\left[\left(1 - 5\alpha^2\right) + i\left(6\alpha - \alpha^3\right)\right]} \tag{5.55}$$

Multiplying and dividing by $(A - iB)$ to Eq. (5.55), where $A = \left(i - 5\alpha^2\right)$, $B = \left(6\alpha - \alpha^3\right)$

$$\beta = \frac{1 \times \left[\left(1 - 5\alpha^2\right) - i\left(6\alpha - \alpha^3\right)\right]}{\left[\left(1 - 5\alpha^2\right) + i\left(6\alpha - \alpha^3\right)\right] \times \left[\left(1 - 5\alpha^2\right) - i\left(6\alpha - \alpha^3\right)\right]} \tag{5.56}$$

$$\beta = \frac{\left(i - 5\alpha^2\right) - i\left(6\alpha - \alpha^3\right)}{\left[\left(1 - 5\alpha^2\right)^2 - i^2\left(6\alpha - \alpha^3\right)^2\right]} \tag{5.57}$$

$$\beta = \frac{\left(1 - 5\alpha^2\right) - i\left(6\alpha - \alpha^3\right)}{\left(1 - 5\alpha^2\right)^2 + \left(6\alpha - \alpha^3\right)^2} \tag{5.58}$$

In order to find out the frequency of oscillator, we assume that the imaginary terms are equal to zero, i.e., $iB = 0$.

$$i\left(6\alpha - \alpha^3\right) = 0 \tag{5.59}$$

$$6\alpha = \alpha^3 \tag{5.60}$$

$$\alpha = \sqrt{6} \tag{5.61}$$

Now, therefore

$$\alpha = \frac{X_{\mathrm{C}}}{R} \tag{5.62}$$

$$\frac{1}{wCR} = \sqrt{6} \tag{5.63}$$

$$w = \frac{1}{\sqrt{6}RC} \tag{5.64}$$

Putting $w = 2\pi f$ in Eq. (5.64), we get

$$2\pi f = \frac{1}{\sqrt{6}RC} \tag{5.65}$$

$$f = \frac{1}{2\pi\sqrt{6}RC} \tag{5.66}$$

- **Value of feedback factor ($\boldsymbol{\beta}$):**

Using Eq. (5.61), put $\alpha = \sqrt{6}$ in Eq. (5.58)

$$\beta = \frac{\left(1 - 5\left(\sqrt{6}\right)^2\right) - i\left(6\left(\sqrt{6}\right) - \left(\sqrt{6}\right)^3\right)}{\left(1 - 5\left(\sqrt{6}\right)^2\right)^2 + \left(6\left(\sqrt{6}\right) - \left(\sqrt{6}\right)^3\right)^2} \Rightarrow \frac{1 - 5 \times 6}{(1 - 5 \times 6)^2} \tag{5.67}$$

$$\beta = -\frac{29}{(-29 \times -29)} \tag{5.68}$$

$$\beta = -\frac{1}{29}(\text{taking Mode at both sides}) \tag{5.69}$$

$$|\beta| = \frac{1}{29} \tag{5.70}$$

Conditions for RC Phase Shift Oscillation

According to Barkhausen criteria, an oscillator will oscillate if and only if $A\beta \geq 1$, putting the value of β from Eq. (5.70), we get

$$A \times \frac{1}{29} \geq 1 \tag{5.71}$$

$$A \geq 29 \tag{5.72}$$

It implies that for RC phase shift oscillator circuit, the amplifier gain (A) should be greater than or equal to 29 in order to provide oscillation.

Question 5.3 In a RC phase shift oscillator, the phase shift network used resistance $R = 4.7\ \text{K}\Omega$, and each capacitor $c = 0.47\ \mu\text{F}$. Find out the frequency of oscillator.

Solution $R = 4.7\ \text{K}\Omega$, and $C = 0.47\ \mu\text{F}$.

$$R = 4.7 \times 10^3\ \Omega C = 0.47 \times 10^{-6}\ \text{F}.$$

Now, the frequency response of the given oscillator is

$$f = \frac{1}{2\pi\sqrt{6}RC} = \frac{1}{2\pi\sqrt{6}\left(4.7 \times 10^3\right)\left(0.47 \times 10^{-6}\right)}$$

$$f = 29.41364\ \text{Hz}.$$

5.6.2 *Lumped or Tuned Oscillator*

- An LC circuit is also called a resonant circuit or tuned circuit consists of inductor and capacitor. Figure 5.4a shows the standard circuitry of tuned circuit where Z_1, Z_2, and Z_3 may be L or C.
- When we connected L and C together, it acts as an electrical resonator.
- LC circuits are used for generating signals at a very high frequency.

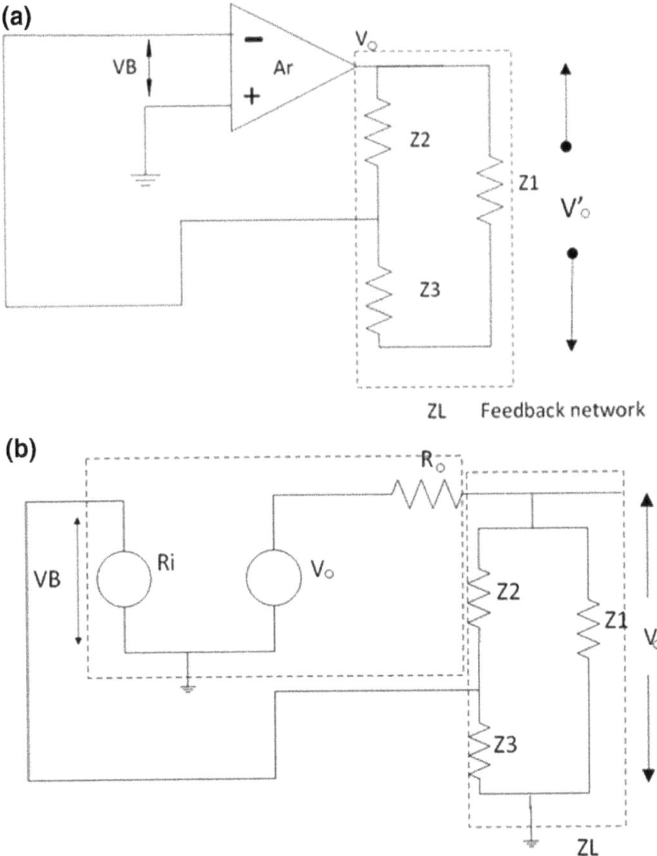

Fig. 5.4 **a** Tuned or Lumped oscillator circuit. **b** Equivalent circuit where Z_1, Z_2, and Z_3 may be L and C

- The examples of tuned oscillators are Hartley oscillator, Colpitts oscillator, and Clapp oscillator, all of these works over high or very high frequency signal.

Figure 5.4b is an equivalent circuit of the tuned circuit in which amplifier is replaced by voltage source and resistor. In Fig. 5.4b, input supply is voltage (V_B) with internal input resistor (R_i) while output terminal carries output voltage (V_O) with internal output resistor (R_O).

Following steps are used to calculate the frequency response of an oscillator.

- **Step 1: Calculating Load Impedance (Z_L)**

Considering Fig. 5.4b, we can say that the load impedance will be

$$Z_L = (Z_2 + Z_3) \| Z_1 \tag{5.73}$$

$$Z_L = \frac{(Z_2 + Z_3) \times Z_1}{Z_2 + Z_3 + Z_1} \tag{5.74}$$

$$Z_L = \frac{Z_1 Z_2 + Z_1 Z_3}{Z_1 + Z_2 + Z_3} \tag{5.75}$$

- **Step 2: Calculate the output voltage (V_0)**

The mathematical formula for calculating the output voltage of the inverting terminal is

$$V_0 = -\frac{R_f}{R_1} V_{in} \tag{5.76}$$

With the help of Fig. 5.4b, the input voltage V_{in} is equal to $V_B A_v$, R_f is equal to Z_L, and R_1 is equal to $(Z_1 + R_0)$.

$$V_0 = -\frac{Z_L}{Z_1 + R_0} V_B A_v \tag{5.77}$$

- **Step 3: The forward gain of oscillator (A)**

Forward gain is the ratio of output voltage to input voltage, where V_0 is output voltage and V_B is input voltage. Using Eq. (5.77), the gain will be

$$A = \frac{\text{output}}{\text{input}} = -\frac{V_0}{V_B} = -\frac{Z_L A_V}{Z_L + R_0} \tag{5.78}$$

- **Step 4: The feedback factor of oscillator (β)**

The value of feedback ratio we can calculate as

$$\beta = -\frac{Z_3}{Z_2 + Z_3} \tag{5.79}$$

- **Step 5: The loop gain of oscillator ($A\ \beta$)**

The loop gain is the product of forward gain (A) and feedback ratio (β), multiplying Eqs. (5.78) and (5.79), we can calculate the value of loop gain ($A\beta$) which is given as

$$A\beta = \frac{Z_L A_V}{(Z_L + R_0)} \times \frac{Z_3}{(Z_2 + Z_3)} \tag{5.80}$$

$$A\beta = \frac{\frac{(Z_1 Z_2 + Z_1 Z_3) A_V}{Z_1 + Z_2 + Z_3}}{\left(\frac{Z_1 Z_2 + Z_1 Z_3}{Z_1 + Z_2 + Z_3} + R_0\right)} \times \frac{Z_3}{Z_2 + Z_3} \tag{5.81}$$

Using Eq. (5.75), put Z_L value in Eq. (5.81), we get

$$A\beta = \frac{\frac{(Z_1Z_2+Z_1Z_3)A_V}{Z_1+Z_2+Z_3}}{\frac{Z_1Z_2+Z_1Z_3+R_0(Z_1+Z_2+Z_3)}{Z_1+Z_2+Z_3}} \times \frac{Z_3}{Z_2+Z_3} \tag{5.82}$$

$$A\beta = \frac{(Z_1Z_2+Z_1Z_3)A_V}{(Z_1Z_2+Z_1Z_3)+(Z_1+Z_2+Z_3)R_O} \times \frac{Z_3}{Z_2+Z_3} \tag{5.83}$$

Now, assuming Z_1, Z_2, and Z_3 are jX_1, jX_2, and jX_3, respectively, then

$$Z_1 = jX_1, \quad Z_2 = jX_2, \quad \text{and} \quad Z_3 = jX_3 \tag{5.84}$$

Using Eqs. (5.83) and (5.84), we get

$$A\beta = \frac{(jX_1.jX_2 + jX_1.jX_3)A_v}{[(jX_1.jX_2)+(jX_1.jX_3)]+(jX_1+jX_2+jX_3)R_O} \times \frac{jX_3}{jX_2+jX_3} \tag{5.85}$$

$$A\beta = \frac{\left(j^2X_1X_2 + j^2X_1X_3\right)A_V}{\left(j^2X_1X_2 + j^2X_1X_2\right) + j(X_1+X_2+X_3)R_O} \times \frac{X_3}{X_2+X_3} \tag{5.86}$$

$$A\beta = \frac{j^2(X_1X_2 + X_1X_3)A_V}{j^2(X_1X_2+X_1X_2)+j(X_1+X_2+X_3)R_O} \times \frac{X_3}{X_2+X_3} \tag{5.87}$$

$$A\beta = \frac{-(X_1X_2 + X_1X_3)A_V}{-(X_1X_2+X_1X_2)+j(X_1+X_2+X_3)R_O} \times \frac{X_3}{X_2+X_3} \tag{5.88}$$

Multiplying and dividing by Eq. (5.88) by $(A - jB)$ where $A = X_1X_2 + X_1X_3$, and $B = X_1 + X_2 + X_3$.

$$A\beta = \frac{-(X_1X_2+X_1X_3)A_V \times [-(X_1X_2+X_1X_2) - j(X_1+X_2+X_3)R_O]}{[-(X_1X_2+X_1X_2)+j(X_1+X_2+X_3)R_O] \times [-(X_1X_2+X_1X_2) - j(X_1+X_2+X_3)R_O]} \times \frac{X_3}{X_2+X_3} \tag{5.89}$$

$$A\beta = \frac{(-X_1X_2+X_1X_3)A_V \times (-(X_1X_2+X_1X_3) - j(X_1+X_2+X_3)R_O)}{(-(X_1X_2+X_1X_3)+j(X_1+X_2+X_3)R_O) \times (-(X_1X_2+X_1X_3) - j(X_1+X_2+X_3)R_O)} \times \frac{X_3}{X_2+X_3} \tag{5.90}$$

$$A\beta = \frac{[(-X_1X_2+X_1X_3)A_V] \times (-(X_1X_2+X_1X_3) - j(X_1+X_2+X_3)R_O)}{\left(-(X_1X_2+X_1X_3)^2+\right) \times \left((X_1+X_2+X_3)^2R_0^2\right)} \times \frac{X_3}{X_2+X_3} \tag{5.91}$$

In order to find out frequency response of the oscillator, we put imaginary part of the Eq. (5.91) equal to zero, i.e., $jB = 0$.

$$j(X_1 + X_2 + X_3)R_O = 0 \tag{5.92}$$

$$X_1 + X_2 + X_3 = 0 \tag{5.93}$$

Equation (5.93) shows the standard mathematical expression of tuned circuit oscillator to calculate frequency response. Using Eq. (5.93), we can calculate the frequency response of the any kind of tuned oscillator such as Hartley, Colpitts, and Clapp oscillator.

5.6.2.1 Hartley Oscillator

The Hartley oscillator is an electronic oscillator that uses an L or C in parallel to determine the frequency. This oscillator is inverted by Ralph Hartley in 1915 (American Scientist). Figure 5.5 shows the circuit diagram of Hartley oscillator.

- **Frequency Response of Hartley Oscillator**:

Using standard frequency response of tuned circuit oscillator from Eq. (5.93), we can say that

$$X_1 + X_2 + X_3 = 0$$

Comparing Figs. 5.5 to 5.4b, X_1 is equal to X_{C1}, X_2 is equal to X_{L1}, and X_3 is equal to X_{L2}, thus replace X_1, X_2, and X_3 by above-mentioned factors, we get

$$\frac{1}{jwC_1} + jwL_1 + jwL_2 = 0 \tag{5.94}$$

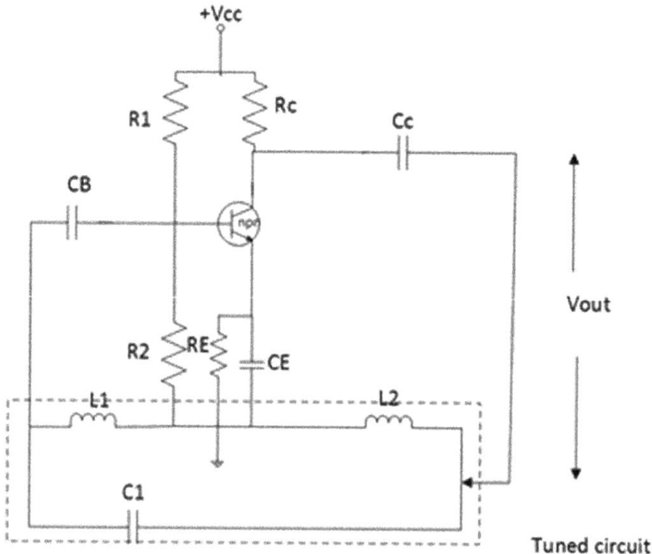

Fig. 5.5 Circuit diagram of Hartley oscillator

$$-\frac{j}{wC_1} + jwL_1 + jwL_2 = 0 \tag{5.95}$$

$$jw(L_1 + L_2) = \frac{j}{wC_1} \tag{5.96}$$

$$w^2 = \frac{1}{C_1(L_1 + L_2)} \tag{5.97}$$

$$w = \sqrt{\frac{1}{C_1(L_1 + L_2)}} \tag{5.98}$$

$$2\pi f = \frac{1}{\sqrt{C_1(L_1 + L_2)}} \tag{5.99}$$

$$f = \frac{1}{2\pi\sqrt{C_1(L_1 + L_2)}} \tag{5.100}$$

Now assuming that $L_1 + L_2 = L_{\text{eq}}$, where is L_{eq} equivalent inductance.

$$\text{Let, } L_1 + L_2 = L_{\text{eq}} \tag{5.101}$$

$$f = \frac{1}{2\pi\sqrt{C_1 L_{\text{eq}}}} \tag{5.102}$$

Equation (5.102) shows the final expression of frequency response for Hartley oscillator.

5.6.2.2 Colpitts Oscillator

- Colpitts oscillator is invented in 1920 by Edurn H. Colpitt.
- It is combination of L and C for frequency determination. Thus, it is also known as LC oscillator.
- The distinguished feature of Colpitts circuit is that the feedback signal is taken from a voltage divider made by two capacitors in series.
- One of the advantages of this circuit is that its implementation is simple and need only a single inductor. Figure 5.6 Circuit diagram of Colpitts oscillator.

- **Frequency Response of Hartley Oscillator**:

Using standard frequency response of tuned circuit oscillator from Eq. (5.93), we can say that

$$X_1 + X_2 + X_3 = 0 \tag{5.93}$$

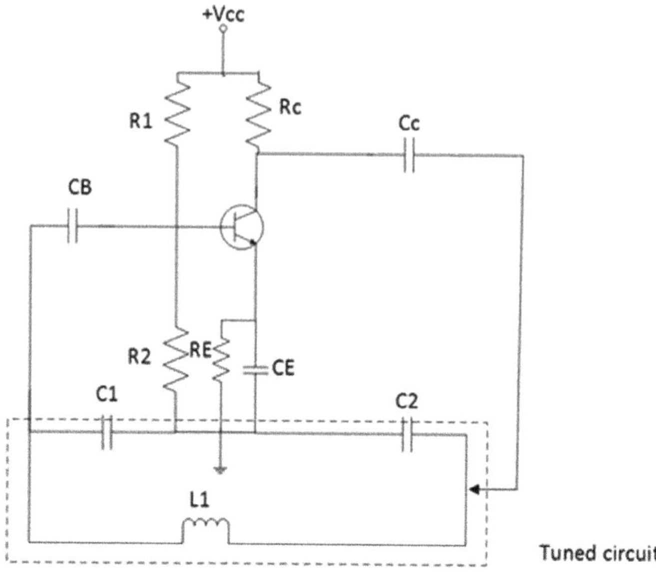

Fig. 5.6 Circuit diagram of Colpitts oscillator

Comparing Figs. 5.6 to 5.4b, X_1 is equal to X_{L1}, X_2 is equal to X_{C1}, and X_3 is equal to X_{C2}, thus replace X_1, X_2, and X_3 by above-mentioned factors, we get

$$jwL_1 + \frac{1}{jwC_1} + \frac{1}{jwC_2} = 0 \tag{5.103}$$

$$jwL_1 - \frac{j}{wC_1} - \frac{j}{wC_2} = 0 \tag{5.104}$$

$$jwL_1 = \frac{j}{wC_1} + \frac{j}{wC_2} \tag{5.105}$$

$$w^2 L_1 = \frac{1}{C_1} + \frac{1}{C_2} \tag{5.106}$$

$$w^2 = \frac{1}{L_1}\left(\frac{C_1 + C_2}{C_1 C_2}\right) \tag{5.107}$$

$$w = \sqrt{\frac{(C_1 + C_2)}{L_1(C_1 C_2)}} \tag{5.108}$$

$$2\pi f = \sqrt{\frac{(C_1 + C_2)}{L_1(C_1 C_2)}} \tag{5.109}$$

$$f = \frac{1}{2\pi}\sqrt{\frac{(C_1 + C_2)}{L_1(C_1C_2)}} \tag{5.100}$$

Now assuming that $\frac{(C_1C_2)}{(C_1+C_2)} = C_{\text{eq}}$, where is C_{eq} equivalent capacitance.

$$\text{Let, } \frac{(C_1C_2)}{(C_1 + C_2)} = C_{\text{eq}} \tag{5.111}$$

$$f = \frac{1}{2\pi\sqrt{LC_{\text{eq}}}} \tag{5.112}$$

Equation (5.112) shows the final expression of frequency response for Colpitts oscillator.

5.6.2.3 Clapp Oscillator

- It is a modified version of Colpitts oscillator.
- The drawback of Colpitts oscillator is that the transistor internal capacitance C_{B} and C_{c} come in parallel with the external capacitance C_1 and C_2, respectively
- Since the internal capacitance C_{B} and C_{c} change due to variation in fixing of Q-point, due to this output frequency of oscillators shows drift to overcome this problem.
- The Colpitts oscillator is modified by Clapp oscillation by adding another capacitance C_0 in series with the inductor L_1 the value of C_1 and C_2 are chosen to be much higher that C_0. Figure 5.7 Circuit diagram of Clapp oscillator.

- **Frequency Response of Hartley Oscillator**:

Using standard frequency response of tuned circuit oscillator from Eq. (5.93), we can say that

$$X_1 + X_2 + X_3 = 0 \tag{5.93}$$

Comparing Fig. 5.7 to 5.4b, X_1 is equal to $(X_{L1} + X_{C0})$, X_2 is equal to X_{C1}, and X_3 is equal to X_{C2}, thus replace X_1, X_2, and X_3 by above-mentioned factors, we get

$$\left(jwL_1 + \frac{1}{jwC_0}\right) + \frac{1}{jwC_1} + \frac{1}{jwC_2} = 0 \tag{5.113}$$

$$jwL_1 - \frac{j}{wC_0} - \frac{j}{wC_1} - \frac{j}{wC_2} = 0 \tag{5.114}$$

$$wL_1 = \frac{j}{wC_0} + \frac{j}{wC_1} + \frac{j}{wC_2} \tag{5.115}$$

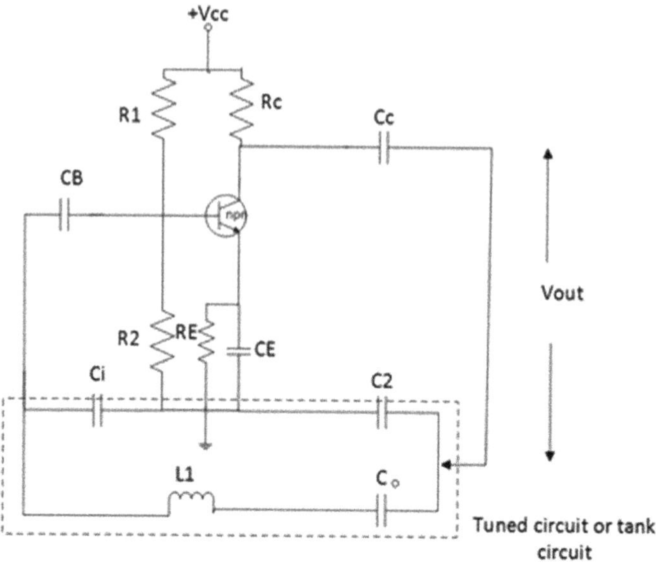

Fig. 5.7 Circuit diagram of Clapp oscillator

$$wL_1 = \frac{1}{w}\left(\frac{1}{C_0} + \frac{1}{C_1} + \frac{1}{C_2}\right) \tag{5.116}$$

$$w^2 = \frac{1}{L_1}\left(\frac{1}{C_0} + \frac{1}{C_1} + \frac{1}{C_2}\right) \tag{5.117}$$

$$w^2 = \frac{1}{L_1}\left(\frac{C_1C_2 + C_0C_2 + C_0C_1}{C_0C_1C_2}\right) \tag{5.118}$$

$$w = \sqrt{\frac{1}{L_1}\left(\frac{C_1C_2 + C_0C_2 + C_0C_1}{C_0C_1C_2}\right)} \tag{5.119}$$

$$2\pi f = \sqrt{\frac{1}{L_1}\left(\frac{C_1C_2 + C_0C_2 + C_0C_1}{C_0C_1C_2}\right)} \tag{5.120}$$

Now assuming that $\frac{C_0C_1C_2}{C_1C_2+C_0C_2+C_0C_1} = C_{\text{eq}}$, where is C_{eq} equivalent capacitance.

$$\text{Let, } C_{\text{eq}} = \frac{C_0C_1C_2}{C_1C_2 + C_0C_2 + C_0C_1} \tag{5.121}$$

$$f = \frac{1}{2\pi\sqrt{L_1C_{\text{eq}}}} \tag{5.122}$$

Equation (5.122) shows the final expression of frequency response for Clapp oscillator.

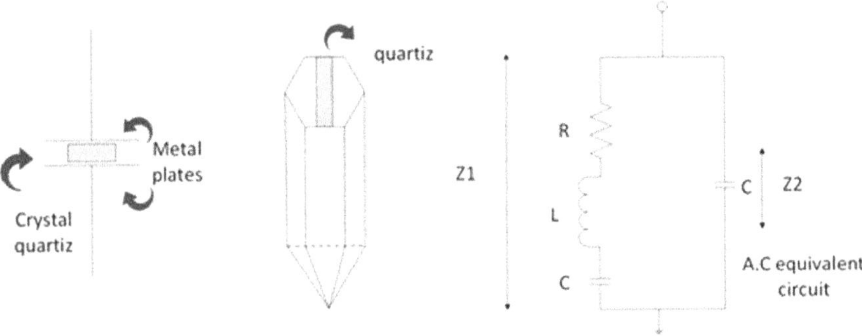

Fig. 5.8 Circuit diagram of crystal oscillator

5.6.3 *Crystal Oscillator*

- The crystal oscillator is an electronic oscillator that uses the mechanical resonance of a vibrating crystal of piezoelectric material to create an electrical signal with a very precise (particular) frequency.
- This frequency is commonly used to provide a stable clock signal to digital integrated circuit and to stabilize frequencies for radio transmitter and receiver.
- The most common type of piezoelectric resonator used in Quartz crystal, so oscillator circuit designed around them becomes known as crystal oscillator. Figure 5.8 Circuit diagram of crystal oscillator.
- Using Fig. 5.8, the resultant impedance (Z) of given circuit is

$$Z = Z_1 \| Z_2 \tag{5.123}$$

where, $Z_1 = R + jwL + \frac{1}{jwC}$, and $Z_2 = X_C$.

Put the value of Z_1 and Z_2 in Eq. (5.123), we get

$$Z = \left(R + jwL + \frac{1}{jwC}\right) || \frac{1}{jwC_0} \tag{5.124}$$

$$Z = \frac{R + jwL + \frac{1}{jwC} \times \frac{1}{jwC_0}}{R + jwL + \frac{1}{jwC} + \frac{1}{jwC_0}} \tag{5.125}$$

- **There are two special cases to calculate frequency response of crystal oscillator**:

Under this subheading, we will discuss about the characteristics of resistor (*R*), i.e., the values of resistor (*R*) are real or imaginary, and analyze what happen if *R* is real or imaginary. Now consider first case when *R* has real value.

Case I: when $R = 0$,
Put $R = 0$ in Eq. (5.125), then

$$Z = \frac{0 + jwL + \frac{1}{jwC} \times \frac{1}{jwC_0}}{0 + jwL + \frac{1}{jwC} + \frac{1}{jwC_0}} \tag{5.126}$$

$$Z = \frac{\left(jwL - \frac{j}{wC}\right) \times \frac{-j}{wC_0}}{\left(jwL - \frac{j}{wC}\right) + \frac{-j}{wC_0}} \tag{5.127}$$

$$Z = \frac{\left(jwL - \frac{j}{wC}\right) \times \frac{-j}{wC_0}}{\left(jwL - \frac{j}{wC}\right) + \frac{-j}{wC_0}} \tag{5.128}$$

$$Z = \frac{j\left(wL - \frac{1}{wC}\right) \times \frac{-1}{wC_0}}{\left(wL - \frac{1}{wC} - \frac{1}{wC_0}\right)} \tag{5.129}$$

$$Z = \frac{-\frac{j}{wC_0}\left(wL - \frac{1}{wC}\right)}{\left(wL - \frac{1}{w}\left(\frac{1}{C} + \frac{1}{C_0}\right)\right)} \tag{5.130}$$

Now we multiply and divide by (w/L) to Eq. (5.130), we get

$$Z = \frac{\left(-\frac{j}{wC_0}\right)\left(\frac{w}{L}\right)\left(wL - \frac{1}{wC}\right)}{\left(\frac{w}{L}\right)\left[wL - \frac{1}{w}\left(\frac{1}{C} + \frac{1}{C_0}\right)\right]} \tag{5.131}$$

$$Z = \frac{-\frac{j}{wC_0}\left(w^2 - \frac{1}{LC}\right)}{\left(w^2 - \frac{1}{L}\left(\frac{1}{C} + \frac{1}{C_0}\right)\right)} \tag{5.132}$$

$$Z = \frac{-\frac{j}{wC_0}\left(w^2 - \frac{1}{LC}\right)}{\left(w^2 - \frac{1}{L}\left(\frac{1}{C} + \frac{1}{C_0}\right)\right)} \tag{5.133}$$

Hence, assuming that

$$w_S^2 = \frac{1}{LC} \tag{5.134}$$

and

$$w_P^2 = \frac{1}{L}\left\{\frac{1}{C} + \frac{1}{C_0}\right\} \tag{5.135}$$

- w_S^2 is series resonant frequency.

- w_P^2 is parallel resonant frequency.

Using Eqs. (5.133), (5.134), and (5.135), we get

$$Z = \frac{-\frac{j}{wC_0}\left(w^2 - w_S^2\right)}{\left(w^2 - w_P^2\right)} \tag{5.1365}$$

Now using Eq. (5.134), we can say that

$$w_s = \frac{1}{\sqrt{LC}} \tag{5.137}$$

$$f_s = \frac{1}{2\pi\sqrt{LC}} \tag{5.138}$$

And, similarly using Eq. (5.135), we can say that

$$w_p = \sqrt{\frac{1}{L}\left\{\frac{1}{C} + \frac{1}{C_0}\right\}} \tag{5.139}$$

$$\text{Let, } C_{eq} = \frac{1}{C} + \frac{1}{C_0} \tag{5.140}$$

$$w_p = \sqrt{\frac{1}{LC_{eq}}} \tag{5.141}$$

$$w_p = \frac{1}{\sqrt{LC_{eq}}} \tag{5.142}$$

$$f_p = \frac{1}{2\pi\sqrt{LC_{eq}}} \tag{5.143}$$

Equations (5.138) and (5.143), both show the frequency response formula for crystal oscillation when resistance is zero, i.e., $R = 0$ in terms of serial resonant frequency and parallel resonant frequency, respectively.

Case II: when $R = jwL$,

Put $R = jwL$ in Eq. (5.125), then

$$Z = \frac{\left(jwL + jwL + \frac{1}{jwC}\right) \times \frac{1}{jwC_0}}{jwL + jwL + \frac{1}{jwC} + \frac{1}{jwC_0}} \tag{5.144}$$

$$Z = \frac{\left(2jwL + \frac{1}{jwC}\right) \times \frac{-j}{wC_0}}{\left(2jwL + \frac{1}{jwC}\right) + \frac{-j}{wC_0}} \tag{5.145}$$

$$Z = \frac{\left(2jwL - \frac{j}{wC}\right) \times \frac{-j}{wC_0}}{\left(2jwL - \frac{j}{wC} - \frac{j}{wC_0}\right)} \tag{5.146}$$

$$Z = \frac{j\left(2wL - \frac{1}{wC}\right) \times \frac{-1}{wC_0}}{\left(2wL - \frac{1}{w}\left(\frac{1}{C} + \frac{1}{C_0}\right)\right)} \tag{5.147}$$

$$Z = \frac{\left(-\frac{j}{wC_0}\right)\left(2wL - \frac{1}{wC}\right)}{\left(2wL - \frac{1}{w}\left(\frac{1}{C} + \frac{1}{C_0}\right)\right)} \tag{5.148}$$

Taking $(2L/w)$ common from numerator and denominator in Eq. (5.148), now the expression will be

$$Z = \frac{\left(-\frac{j}{wC_0}\right)\left(\frac{2L}{w}\right)\left(w^2 - \frac{1}{wC\left(\frac{2L}{w}\right)}\right)}{\left(\frac{2L}{w}\right)\left(w^2 - \frac{1}{2L}\left(\frac{1}{C} + \frac{1}{C_0}\right)\right)} \tag{5.149}$$

$$Z = \frac{\left(-\frac{j}{wC_0}\right)\left(w^2 - \frac{1}{2LC}\right)}{\left(w^2 - \frac{1}{2L}\left(\frac{1}{C} + \frac{1}{C_0}\right)\right)} \tag{5.150}$$

Hence, assuming that

$$w_S^2 = \frac{1}{2LC} \tag{5.151}$$

and

$$w_P^2 = \frac{1}{2L\left(\frac{1}{C} + \frac{1}{C_0}\right)} \tag{5.152}$$

Using Eqs. (5.150), (5.151), and (5.152), we get

$$Z = \frac{\left(-\frac{j}{wC_0}\right)\left(w^2 - w_S^2\right)}{\left(w^2 - w_P^2\right)} \tag{5.153}$$

Using Eq. (5.151), we can say that

$$w_s = \sqrt{\frac{1}{2LC}} \tag{5.154}$$

$$w_s = \frac{1}{\sqrt{2LC}} \tag{5.155}$$

$$f_s = \frac{1}{2\pi\sqrt{2LC}} \tag{5.156}$$

Similarly, using Eq. (5.152), we can say that

$$w_p = \sqrt{\frac{1}{2L}\left(\frac{1}{C} + \frac{1}{C_0}\right)} \tag{5.157}$$

$$w_p = \sqrt{\frac{1}{2L\left(\frac{C \cdot C_0}{C+C_0}\right)}} \tag{5.158}$$

$$\text{Let } C_{eq} = \frac{C \cdot C_0}{C + C_0} \tag{5.159}$$

$$w_p = \frac{1}{\sqrt{2LC_{eq}}} \tag{5.160}$$

$$\text{So, } f_p = \frac{1}{2\pi\sqrt{2LC_{eq}}} \tag{5.161}$$

Equations (5.156) and (5.161), both show the frequency response formula for crystal oscillation when resistance is zero, i.e., $R = jwL$ in terms of serial resonant frequency and parallel resonant frequency, respectively.

5.7 Quality of Factor

- Due to high quality of factor of a resonant frequency, it provides very good frequency stability.
- It is denoted by "Q".
- The quality of factor "Q" is very high, typically 20,000 and up to 10^6 can be achieved.
- There is a mathematical formula to calculate quality of factor given as

$$Q = \frac{wL}{R} \tag{5.162}$$

In Eq. (5.162), Q shows quality of factor, it should be as large as possible. w shows angular frequency, and $w = 2\pi f$ where f is frequency components, L is inductor, and R is resistor of the given circuit.

Question 5.4 In a crystal oscillator $L = 0.4$ H, $C = 0.085$ pF and $C_0 = 1$ pF with $R = 5$ KΩ Calculate

(i) Series resonant frequency (W_S).

(ii) Parallel resonant frequency (W_P).
(iii) Quality factor (Q).
(iv) By what percentage quality factor will improve.

Solution $L = 0.4$ H, $C = 0.85$ pF, $C_0 = 1$ pF, and $R = 5$ KΩ.

(i) The series resonant frequency will be

$$f_s = \frac{1}{2\pi\sqrt{LC}}$$
$$f_s = \frac{1}{2\pi\sqrt{0.4 \times 0.85 \times 10^{-12}}}$$
$$f_s = 0.863138\,\text{MHz}.$$

(ii) The parallel resonant frequency will be

$$f_p = \frac{1}{2\pi\sqrt{LC_{eq}}}$$
$$C_{eq} = c \cdot \frac{c_0}{c + c_0}$$
$$= \frac{0.085 \times 10^{-12} \times 1 \times 10^{-12}}{0.085 \times 10^{-12} + 1 \times 10^{-12}}$$
$$= 7.834101 \times 10^{-14} F$$
$$\Rightarrow \frac{f_p 1}{2\pi\sqrt{0.4 \times 7.834101 \times 10^{-14}}}$$
$$\Rightarrow 0.899074\,\text{MHz}.$$

(iii) The quality factor will be

$$Q = \frac{WL}{R}$$
$$= \frac{2\pi \times f_s \times 0.4}{5 \times 10^3}$$
$$= \frac{2\pi 0.863138 \times 10^6 \times 0.4}{5 \times 10^3}$$
$$= 433.860.$$

(iv) The percentage of quality factor improving is

$$Q = \frac{f_p - f_s}{f_p} \times 100$$
$$= \frac{0.89974 \times 10^6 - 0.86138 \times 10^6}{0.899074 \times 10^3} \times 100$$

$$= 3.997\%.$$

Question 5.5 In crystal oscillator $L = 50$ mH, $R = 500\ \Omega$, $C = 0.02$ pF, and $C_0 = 12$ pF, Compute serial resonant frequency (f_s) and parallel resonant frequency (f_p). If the external capacitance across the crystal changes from 5 to 6 pF. Calculate the change in frequency of oscillators.

Solution $L = 50\,\text{mH} \Rightarrow 50 \times 10^{-3}$ H, $R = 500\,\Omega$, $C = 0.02\,\text{pF} \Rightarrow 0.02 \times 10^{-12}$ F, $C_0 = 12\,\text{pF}$,

(i) Serial resonant frequency

$$f_s = \frac{1}{2\pi\sqrt{LC}}$$
$$f_s = \frac{1}{2\pi\sqrt{50 \times 10^{-3} \times 0.2 \times 10^{-12}}}$$
$$f_s = 5.0329\,\text{MHz}.$$

(ii) Parallel resonant frequency

$$f_p = \frac{1}{2\pi\sqrt{LC_{eq}}}$$
$$C_{eq} = \frac{c.c_0}{c + c_0} \Rightarrow \frac{0.02 \times 10^{-12} \times 12 \times 10^{-12}}{0.02 \times 10^{-12} + 12 \times 10^{-12}}$$
$$C_{eq} = 0.0199667 \times 10^{-12}\,\text{F}$$
$$f_p = \frac{1}{2\pi\sqrt{50 \times 10^{-3} \times 0.0199667 \times 10^{-12}}}$$
$$f_p = 5.037116\,\text{MHz}.$$

(iii) If capacitor varies 5 to 6 pF, then case I if $C' = 5\,\text{pF}$, parallel resonant frequency

$$y = C_0 + C'$$
$$y' = 12 \times 10^{-12} + 5 \times 10^{-12}$$
$$y' = 17 \times 10^{-12}\,\text{F}$$
$$C_{eq} = \frac{c \cdot y}{c + y} = \frac{0.02 \times 10^{-12} \times 17 \times 10^{-12}}{0.02 \times 10^{-12} + 17 \times 10^{-12}}$$
$$C_{eq} = 0.019976498 \times 10^{-12}\,\text{f},$$
$$\text{And, } f_p' = \frac{1}{2\pi\sqrt{LC_{eq}}} \Rightarrow \frac{1}{2\pi\sqrt{50 \times 10^{-3} \times 0.019976498 \times 10^{-12}}}$$
$$f_p' = 5.03588\,\text{MHz}.$$

(iv) Case II if $C' = 6\,\text{pF}$, parallel resonant frequency

$$y' = C_0 + C''$$
$$y' = 12 \times 10^{-12} + 6 \times 10^{-12}$$
$$y' = 18 \times 10^{-12} F$$
$$C_{\text{eq}} = \frac{c \cdot y'}{c + y'} = \frac{0.02 \times 10^{-12} \times 18 \times 10^{-12}}{0.02 \times 10^{-12} + 18 \times 10^{-12}}$$
$$C_{\text{eq}} = 0.19977802 \times 10^{-12}$$
$$f_{\text{p}}'' = \frac{1}{2\pi\sqrt{LC_{\text{eq}}}} \Rightarrow \frac{1}{2\pi\sqrt{50 \times 10^{-3} \times 0.019977802 \times 10^{-12}}}$$
$$f_{\text{p}}'' = 5.035716\,\text{MHz}.$$

(v) Finally, the change in frequency of oscillators

$$f_{\text{change}} = f_{\text{p}}' - f_{\text{p}}'' \Rightarrow 5.03588 - 5.03716$$
$$f_{\text{change}} = 1.64 \times^{10^{-4}}\ \text{MHz}$$
$$f_{\text{change}} = 1.64 \times 10^{-4} \times 10^{6}$$
$$f_{\text{change}} = 164\,\text{Hz}.$$

Question 5.6 $L = 0.1$ H, $C = 0.1$ pF, $R = 10\,\text{K}\Omega$, and $C_0 = 1$ pF. Calculate f_{s} and Q.

Solutions $L = 0.1\,\text{H}$, $C = 0.1\,\text{pF} \Rightarrow 0.01 \times 10^{-12}\,\text{F}$, $R = 10\,\text{K}\Omega \Rightarrow 10 \times 10^{3}\Omega$, $C_0 = 1\,\text{pF} \Rightarrow 1 \times 10^{-12}\,\text{F}$,

(i) The serial resonant frequency will be

$$f_{\text{s}} = \frac{1}{2\pi\sqrt{LC}} \Rightarrow \frac{1}{2\pi\sqrt{0.1 \times 0.01 \times 10^{-12}}}$$
$$f_{\text{s}} = 5.032921\,\text{MHz}.$$

(ii) The quality factor will be

$$Q = \frac{wL}{R} \Rightarrow \frac{2\pi f_{\text{s}} \times L}{R}$$
$$Q = \frac{2\pi \times 5.032921 \times 0.1}{10 \times 10^{3}}$$
$$Q = 0.31622 \times 10^{3}$$
$$Q = 316.22.$$

Questions 5.7 In Colpitts oscillator $C_1 = 150\,\text{pF}$, $C_2 = 1.5\,\text{nF}$, and $L = 50\,\mu\text{H}$. Calculate frequency of the system.

Solution $C_1 = 150\,\text{pF} \Rightarrow 150 \times 10^{-12}$, $C_2 = 1.5\,\text{nF} \Rightarrow 1.5 \times 10^{-9}\,\text{F}$, $L = 50\,\mu\text{H} \Rightarrow 50 \times 10^{-6}\,\text{H}$,

(i) Frequency of Colpitts oscillator is

$$f = \frac{1}{2\pi\sqrt{LC_{\text{eq}}}}$$

$$C_{\text{eq}} = \frac{c_1 c_2}{c_1 + c_2} \Rightarrow \frac{150 \times 10^{-12} \times 1.5 \times 10^{-9}}{150 \times 10^{-12} + 1.5 \times 10^{-9}}$$

$$C_{\text{eq}} = \frac{2.25 \times 10^{-19}}{1.65 \times 10^{-9}}$$

$$C_{\text{eq}} = 136.36364 \times 10^{-12}\,\text{F}.$$

$$f = \frac{1}{2\pi\left(\sqrt{50 \times 10^{-6} \times 136.36364 \times 10^{-12}}\right)}$$

$$f = 1.927462\,\text{MHz}.$$

Question 5.8 In a Clapp oscillator value of $C_1 = C_2 = 0.001\,\mu\text{F}$, $L = 15\,\mu\text{H}$, and $C_3 = 50\,\text{pF}$. Compute the oscillator frequency.

Solution $C_1 = C_2 = 0.001\,\mu\text{F} \Rightarrow 0.001 \times 10^{-6}$, $C_3 = 50\,\text{pF} \Rightarrow 50 \times 10^{-12}$, F, $L = 15\,\mu\text{H} \Rightarrow 15 \times 10^{-6}\,\text{H}$,

(i) Frequency of Clapp oscillator is

$$f = \frac{1}{2\pi\sqrt{LC_{\text{eq}}}}$$

$$C_{\text{eq}} = \frac{C_0 C_1 C_2}{C_1 C_2 + C_0 C_2 + C_0 C_1}$$

$$C_{\text{eq}} = \frac{\left(0.001 \times 10^{-6}\right)\left(50 \times 10^{-12}\right)}{(0.001 \times 10^{-6})^2 + 0.001 \times 10^{-6} \times 50 \times 10^{-12} + (50 \times 10^{-12} \times 0.001 \times 10^{-6})}$$

$$C_{\text{eq}} = \frac{\left(5 \times 10^{-23}\right)}{1.1 \times 10^{-12}} \Rightarrow 4.545454 \times 10^{-11}$$

$$C_{\text{eq}} = 45.4545 \times 10^{-12}\,\text{F}.$$

$$f = \frac{1}{2\pi\sqrt{LC_{\text{eq}}}} \Rightarrow \frac{1}{2\pi\sqrt{15 \times 10^{-6} \times 45.4545 \times 10^{-12}}}$$

$$f = 6.095173\,\text{MHz}.$$

Question 5.9 If 63 pF capacitor is connected in series with 5 μH inductor. Find out frequency of Clapp oscillator.

Solution $C_3 = 63\,\text{pF}$, $C_{eq} = C_3 = 63 \times 10^{12}\,\text{F}$, $L = 5\,\mu\text{H} \Rightarrow 5 \times 10^{-6}$.

(i) Frequency of Clapp oscillator is

$$f = \frac{1}{2\pi\sqrt{LC_{eq}}} \Rightarrow \frac{1}{2\pi\left(\sqrt{63 \times 10^{-12} \times 10^{-6}}\right)}$$

$$f = 8.967365\,\text{MHz}.$$

Question 5.10 In the given Hartley oscillator, calculate L_2 if $L_1 = 15\,\text{mH}$, $c = 50\,\text{pF}$, mutual inductance $= 5\,\mu\text{F}$, and $f = 168\,\text{KHz}$.

Solution $L_1 = 15\,\text{mH} \Rightarrow 15 \times 10^3$, $C = 50\,\text{pF} \Rightarrow 50 \times 10^{-12}$, $M = 5\,\mu\text{F} \Rightarrow 5 \times 10^{-6}\,\text{F}$, $f = 168\,\text{KHz} \Rightarrow 168 \times 10^3\,\text{Hz}$,

(i) In Hartley oscillator, the value of L_2 will be,

$$f = \frac{1}{2\pi\sqrt{L_{eq}C}}$$

$$\text{here } L_{eq} = L_1 + L_2 + 2M$$

$$L_{eq} = 15 \times 10^3 + L_2 + 2 \times 5 \times 10^{-6}$$

$$L_{eq} = 0.1501 + L_2$$

$$168 \times 10^3 = \frac{1}{2\pi\sqrt{(0.1501 + L_2)\left(50 \times 10^{-12}\right)}}$$

$$\left(168 \times 10^3\right)^2 = \frac{1}{(2\pi)^2(0.1501 + L_2)\left(50 \times 10^{-12}\right)}$$

$$0.1501 + L_2 = \frac{1}{4\pi r^2 \times \left(168 \times 10^3\right)^2 \times \left(50 \times 10^{-12}\right)}$$

$$0.1501 + L_2 = 0.017949472$$

$$L_2 = 0.017949472 - 0.01501$$

$$L_2 = 2.9394 \times 10^{-3}\,\text{H}.$$

Question 5.11 In Hartley oscillator, two inductor $L_1 = L_2 = 2\,\text{mH}$ and frequency varies from 950 to 2050 kHz. Find the range of capacitor.

Solution $L_1 = L_2 = 2\,\text{mH} \Rightarrow 2 \times 10^{-3}\,\text{H}$,

(i) In Hartley oscillator, the range of capacitor will be

$$f = \frac{1}{2\pi\left(\sqrt{L_{eq}C}\right)}$$

$$L_{eq} = L_1 + L_2 = \left(2 \times 10^{-3}\right) + \left(2 \times 10^{-3}\right)\text{H} \Rightarrow 4 \times 10^{-3}\,\text{H}.$$

(a) At $f = 950$ kHz,

$$950 \times 10^3 = \frac{1}{2\pi\sqrt{4 \times 10^{-3} \times C}}$$

$$C = \frac{1}{(2\pi)^2\left(950 \times 10^3\right)^2\left(4 \times 10^{-3}\right)}$$

$$C = \left(7.024 \times 10^{-12}\right)\text{F}.$$

(b) At f =2050 kHz,

$$2050 \times 10^3 = \frac{1}{2\pi\sqrt{4 \times 10^{-3} \times C}}$$

$$C = \frac{1}{(2\pi)^2\left(2050 \times 10^3\right)^2\left(4 \times 10^{-3}\right)}$$

$$C = \left(1.508 \times 10^{-12}\right)\text{F}.$$

Summary

- **This chapter included introduction of oscillator.**
- **Barkhausen criteria for oscillation.**
- **Classification of oscillator.**
- **Analyze the frequency response of Wien-bridge, and RC phase shift.**
- **Discuss lumped oscillators, i.e., Hartley, Clapp, and Colpitts.**
- **Define crystal oscillator.**
- **Define quality of factor (QoS) of the system.**

Numerical Problems

Question 1: Explain Barkhausen criterion and stability problem.
Question 2: Discuss the Hartley oscillator and show its frequency response formula.
Question 3: In Hartley oscillator two inductor $L_1 = 10$ mH, $L_2 = 5$ mH and frequency is 1850 kHz. Compute the value of capacitor.
Question 4: Explain crystal oscillator.
Question 5: Explain Wien-bridge oscillator with frequency response formula.
Question 6: In crystal oscillator $R = 100\,\Omega$, $C = 0.6$ pF, $L = 15$ mH, and $C_0 = 8$ pF. Compute serial resonant frequency (f_s). If the external capacitance across the crystal changes and becomes 4 pF. Calculate the new frequency of oscillators.

Question 7: $C_1 = 150\,\text{pF}$, $C_2 = 1.5\,\text{nF}$, $L = 50\,\mu\text{H}$. Calculate the frequency for Colpitts oscillator.

Question 8: In Wien-bridge oscillator $R = 5.1\,\text{K}\Omega$ $C = 1\,\text{nF}$ $R_3 = 12\,\text{K}\Omega$, $R_4 = 5.1\,\text{K}\Omega$

(a) Whether this oscillator oscillates or not.
(b) Calculate oscillator output frequency.

Chapter 6
OP-AMP Applications, Timer, Voltage Regulator, and Converter

Learning Objectives

- **Introduce multivibrator and their classifications.**
- **Discuss A-stable, Mono-stable, and Bi-stable multivibrators.**
- **Define Schmitt trigger, VCO, and PLL.**
- **Discuss filters with their frequency response.**
- **Define voltage regulator and its types.**
- **Discuss fixed and adjustable voltage regulator.**
- **Introduce sample and hold circuit, A to D, and D to A converter.**

6.1 Multivibrator

Multivibrator is an electronic circuit which can have one or two stable states of operation. Depending on the number of stable, we can classify the multivibrator as shown below.

6.2 Classifications of Multivibrator

1. A-stable multivibrator.
2. Mono-stable multivibrator.
3. Bi-stable multivibrator.

G. S. Tomar and A. Bagwari, *Fundamentals of Electronic Devices and Circuits*, Algorithms for Intelligent Systems, https://doi.org/10.1007/978-981-15-0267-5_6

(Multivibrator)

(A − stable Multivibrator) (Mono − stable Multivibrator) (Bi − stable Multivibrator)

6.2.1 A-stable Multivibrator

A multivibrator is an electronic circuit used to implement a variety of simple two-state systems such as oscillators, timers, and flip-flop. As A-stable multivibrator has two states neither one stable. The circuit, therefore, behaves as an oscillator with the time spent in each state controlled by the charging or discharging of a capacitor "*C*" through a resistor.

The A-stable multivibrator may be created directly with transistor or with use of (IC) such as OP-AMP or the 555 timer.

Most OP-AMP are powered by positive and negative rail voltage; the output never able to exceed these rail voltages. Depending upon initial conditions, the OP-AMP output will drive to either positive or negative rail. Upon this occurrence, the capacitor will either charge or discharge through the resistor "R_2", i.e., voltage slowly rising or falling.

As soon as the voltage at the OP-AMP inverting terminal reaches that at the non-inverting terminal (the OP-AMP output voltage divided by R_1 and R_2), the output will drive to the opposing rail and this process will repeat with the capacitor discharging if it had previously charged and vice versa. Once the inverting terminal reaches the voltage of the non-inverting terminal, the output again drives to the opposing rail voltage and the cycle beings again. Thus, the A-stable multivibrator creates a square wave with no inputs.

In A-stable multivibrator, output does not have any stable state. Its output changes its state from high to low and low to high repeatedly. A-stable multivibrator has two quasi-stable states and no-stable state. Its output changes its state from one quasi-state to the other. It does not require any external trigger input to change its state. Hence, it is also called a "free-running multivibrator". The output voltage waveform of an A-stable multivibrator is shown in Fig. 6.1. It is a rectangular waveform. Period of A-stable multivibrator displayed in Fig. 6.2, $T = 0.693\ C$.

$$T = 2 \times C_1 \times R_3 \times \log_e\left[1 + \frac{2 \times R_2}{R_1}\right] \quad (6.1)$$

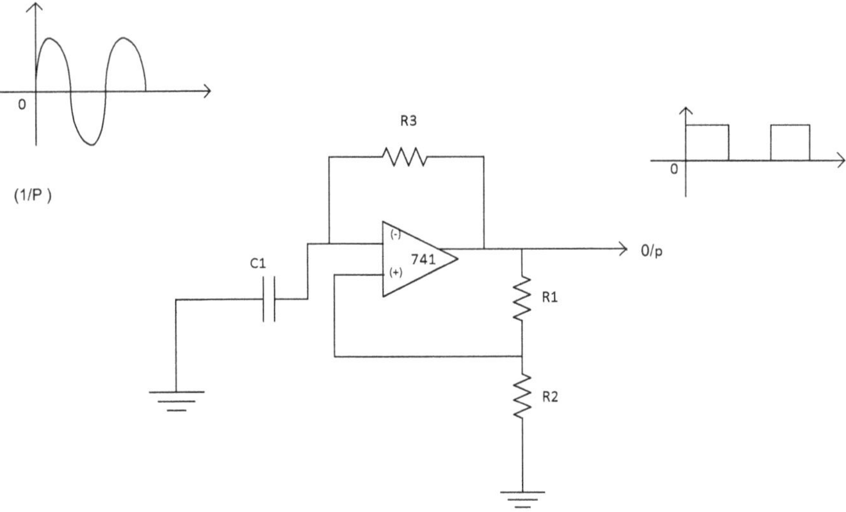

Fig. 6.1 A-stable multivibrator

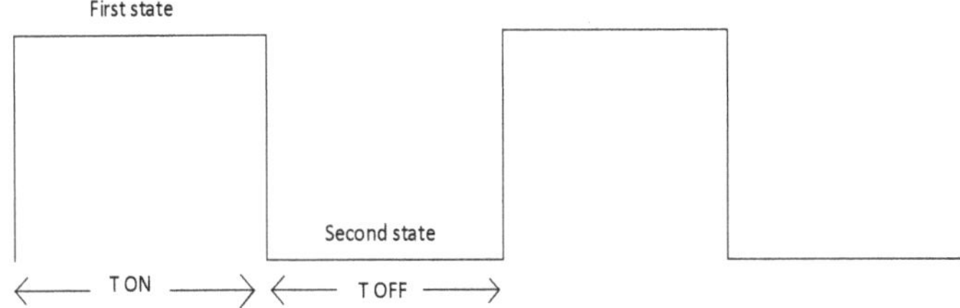

Fig. 6.2 Output of A-stable multivibrator

6.2.1.1 Application of A-stable Multivibrator

(i) To obtain a perfectly square waveform at the output of an A-stable multivibrator, total time period of one cycle of the output is given by.

$$T = T_{\mathrm{ON}} + T_{\mathrm{OFF}} \tag{6.2}$$

$$f = \frac{1}{T} \tag{6.3}$$

(ii) To generate rectangular wave.
(iii) As a square wave generator.
(iv) In the ramp generator.
(v) In the flasher circuit.

6.2.2 *Mono-stable Multivibrator*

Mono-stable multivibrator has one stable state and another quasi-stable state. It requires an external trigger input to change its state from stable to quasi-stable. After some time, it will automatically return back to the stable state. The duration for which output remains in the quasi-stable state is called “pulse-width”.

Figure 6.3 illustrates a mono-stable multivibrator circuit that employs a single OP-AMP. The main component of the circuit is the 741, a general purpose operational amplifier. A mono-stable multivibrator is a timing circuit that changes its state once triggered but returns to its original state after a certain time delay. It got its name from the fact that only one of its output states is “stable”, and it is known as “one-shot”.

A negative trigger pulse at the input forces the output to the OP-AMP to logic “high”. This charges up capacitor “C_2” which keeps the non-inverting input of the OP-AMP temporarily higher than the inverting input, maintaining the output high for a certain period of time, eventually capacitor “C_2” discharges to grand and the OP-AMP output swings back to logic “low”. The duration of the pulse is defined by R_2 and C_2. The one-shot has several applications which include dividing the frequency of input signal and converting an irregular input pulse to a uniform output pulse. The formula to calculate the period width of mono-stable multivibrator can be written as

Pulse width (T) = 0.69 RC

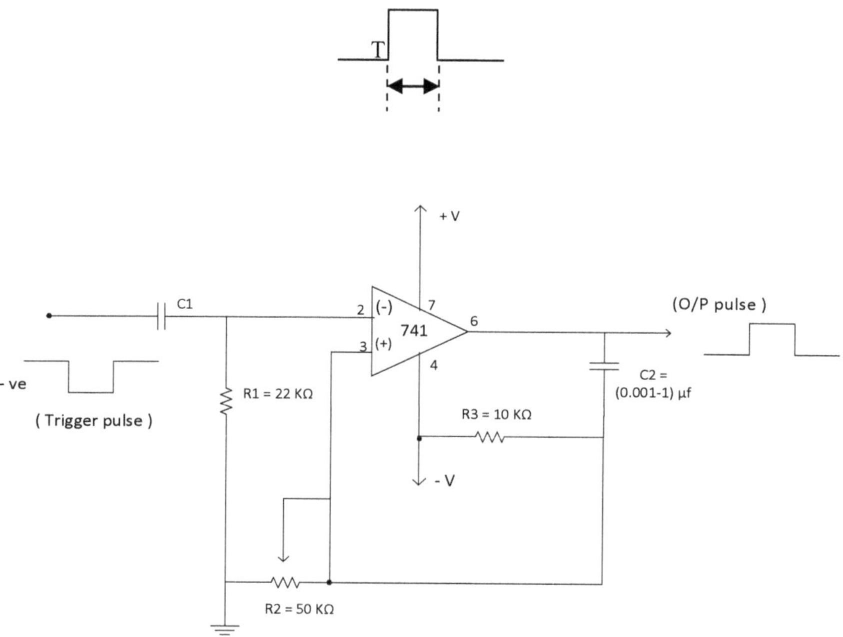

Fig. 6.3 Mono-stable multivibrator

6.2.2.1 Application of Mono-stable Multivibrator

1. As a delay generator.
2. Used to gate another circuit.
3. As a timer.

6.2.2.2 Drawback of Mono-stable Multivibrator

1. Tolerance of the value of resistor "R" and capacitor "C".
2. Temperature change.
3. Change in the transmitter parameters (external pulse needed to change the stable state of the circuit).

6.2.3 *Bi-stable Multivibrator*

A bi-stable multivibrator has two stable states. It needs external trigger input to change the existing stage. Bi-stable multivibrator is basically a flip-flop. The bi-stable multivibrator has no quasi-stable state. Figure 6.4 shows the output waveform of bi-stable multivibrator.

Working of bi-stable multivibrator: In the given Fig. 6.4, assuming that both transmitters (Q_1 and Q_2) are identical. When transistor Q_1 is conducting $\overparen{\text{(ON)}}$, its respective output Q becomes high and Q becomes low and the transistor Q_2 is OFF.

To make Q low, we apply negative trigger pulse. Q_1 will be out of conduction (OFF). Its collector voltage increases to ($+V_{CC}$). Due to cut-off, the collector voltage of Q_1 will increase to $+V_{CC}$ and Q_2 will be conducted (ON). Output of Q is low and $\overline{Q}$ will be high.

6.3 Comparative Study between All Multivibrators

Comparative Study between all Multivibrators

S. No.	A-stable	Mono-stable	Bi-stable
1.	There are no-stable states of output	There is only one stable state of the output	There are two stable states of the output
2.	Trigger input is not necessary for changing the state of the output	Trigger pulse is required for changing the state of output	Trigger input is required for charging the state of output

(continued)

(continued)

S. No.	A-stable	Mono-stable	Bi-stable
3.	Number of quasi-stable state is two	Number of quasi-stable states is one	No quasi-stable states
4.	Used as rectangular, square wave, or ramp generator	Used as a timer	Used as flip-flop
5.	Time for the two quasi-stable states depends on RC-time constant. The two quasi-stable states can have different intervals $T = 0.693$ RC	Time for the quasi-stable state depends on RC-time constant $T = 0.69$ RC	No quasi-stable states

6.4 Schmitt Trigger

This circuit is a fast operating voltage level detector. When the V_{in} arrives at the upper or lower trigger level, the output charges rapidly and the circuit operates with almost any type of input wave form, and gives a "pulse-type output". The circuit of an OP-AMP Schmitt-trigger circuit is shown in Fig. 6.5.

The V_{in} is applied to the inverting input terminal, and the feedback voltage goes to the non-inverting terminal. This means that the circuit uses positive voltage feedback instead of negative feedback, i.e., in this circuit, feedback voltage aids the V_{in} rather than opposing it, for instance, assume the inverting V_{in} to be slightly positive. This will produce a negative V_{out}. The divider feeds a negative voltage to the non-inverting input, which results a large negative voltage. This feedback shows more negative voltage until the circuit is driven into negative saturation. If the V_{in} were slightly negative instead of positive, the circuit would be driven into the positive saturation. This is the reason why the circuit is also referred to as "regenerative comparator".

When the circuit is positively saturated, a positive voltage is feedback to the non-inverting input. This positive input holds the output in the high state. Similarly, when the output voltage V_{out} is negatively saturated, a negative "Voltage" is fed to the non-inverting input which holding the output in the low state. In either case, the positive feedback reinforces the existing output state.

$$\text{The feedback fraction } \beta = \left(\frac{R_2}{R_1} + R_2\right) \tag{6.4}$$

1. When output is positively saturated, the reference voltage applied to the non-inverting input is

$$V_{\text{ref}} = +\beta \times V_{\text{saturation}} \tag{6.5}$$

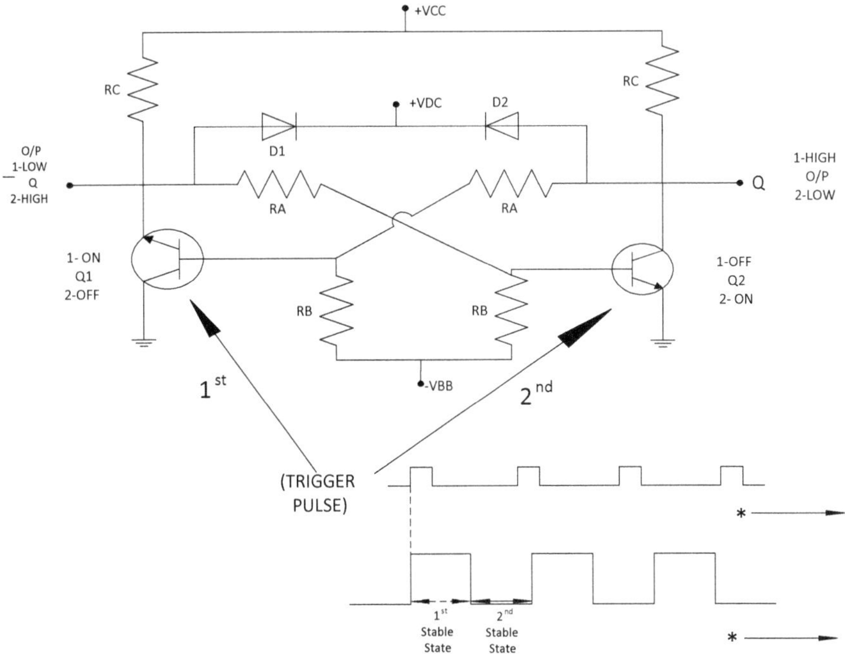

Fig. 6.4 Bi-stable multivibrator

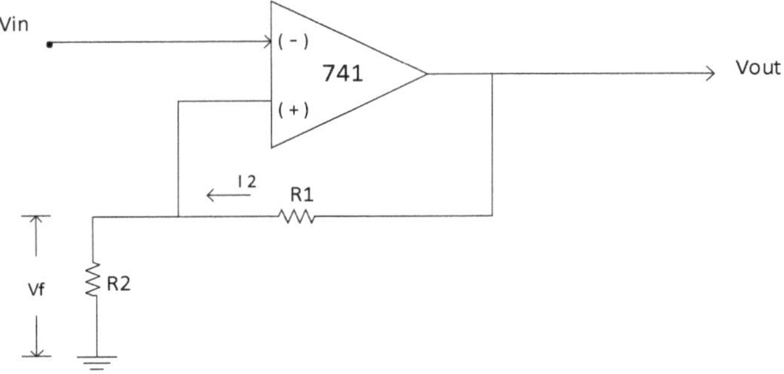

Fig. 6.5 Schmitt trigger

2. When output is negatively saturated, the

$$V_{\text{ref}} = -\beta \times V_{\text{saturation}} \tag{6.6}$$

6.5 VCO and PLL

6.5.1 VCO or Voltage Control Oscillator (Voltage to Frequency Converter)

VCO is a circuit which produces oscillation where frequency is controlled by voltage. Hence, it is named as "voltage control oscillator". Most frequently used VCO is NE/SE 566. Figure 6.6 shows the circuitry of voltage control oscillator (VCO), whereas Fig. 6.7 depicts the IC configuration of VCO.

6.5.1.1 Working of VCO

Considering Fig. 6.8, the frequency of oscillation depends upon external resistance (R_1), C_1, and V_C applied at input terminal (5). The output of VCO can be a triangular wave and a square wave. The triangular wave generated by alternatively charging "C_1" via current source & linearly discharging then. The charging and discharging level of "C" is determined by "Schmitt trigger".

As Schmitt trigger generates a "square wave", its output can be direct taken by pin no. (3). Both waveforms are buffered so that output impedance of each is 50 Ω. The amplitude of triangular wave should be "2.4 V" (peak-peak) for triangular wave and "5.4 V" (peak-peak) for square wave.

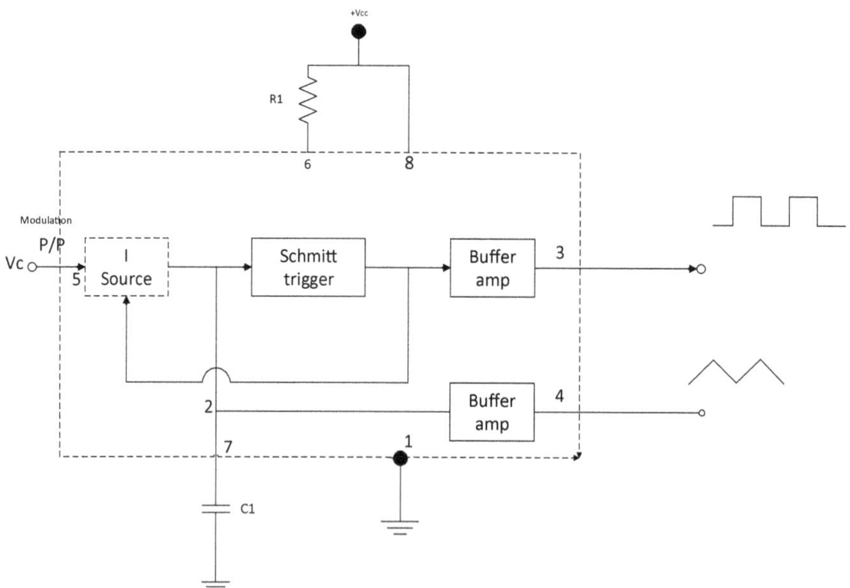

Fig. 6.6 Voltage control oscillator

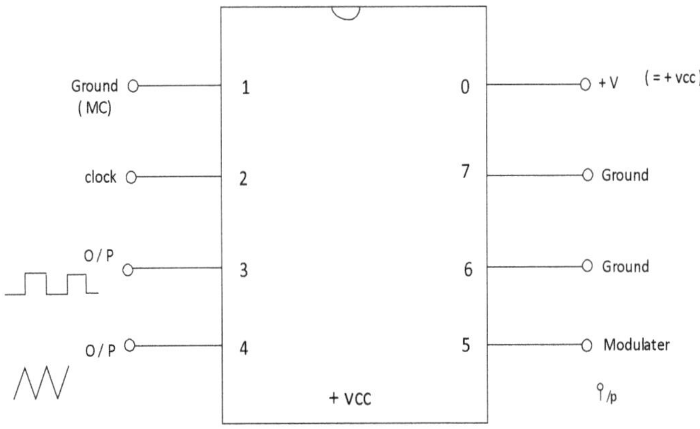

Fig. 6.7 IC configuration of VCO

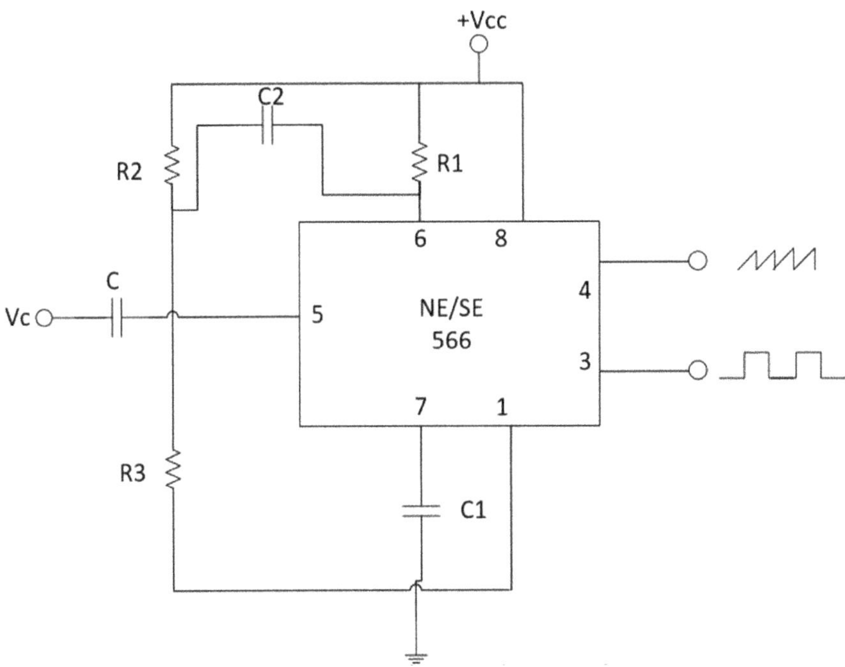

Fig. 6.8 External circuit diagram of VCO

In a circuit (R_1, C_1) determine free-running frequency, at potential "V_C" at terminal (5), which is formed by voltage divider arrangement R_1 and R_3. Also help in setting-up oscillating frequency, voltage "V_C" at Pin-5 should be in range of [3/4(+v) $\leq V_C \leq$ (+v)].

Where (+)V is positive supply voltage. Now, the resonance frequency of oscillation is

$$f_0 = \frac{2 \times [+V - V_C]}{R_1 \times C_1 \times (+V)} \tag{6.7}$$

In the equation (6.7), R_1 lies between (2 kW to 20 kW), input modulated signal is applied through capacitor "C", small capacitor "C_2" is connected between (Pin 5 & 6), and the maximum frequency at the output of VCO is 1 MHz.

6.5.1.2 Application of VCO

VCO is commonly used for converting low frequency signals into Audio-frequency signal. The low frequency signals are electro cardiogram (ECG) or electro encepelogram (EEG). These low-frequency signals can be transmitted by a transmitter line for digenesis purpose or can be stored in magnetic disk or tape only when they are converted into audio-frequency range.

6.5.2 *Phase Locked Loop (PLL)*

It is used to trace the incoming signal with respect to its phase angle and frequency. Figure 6.9 shows the block diagram PLL, which consists of

1. Phase detector.
2. Low pass filter (LPF).
3. Amplifier.
4. VCO

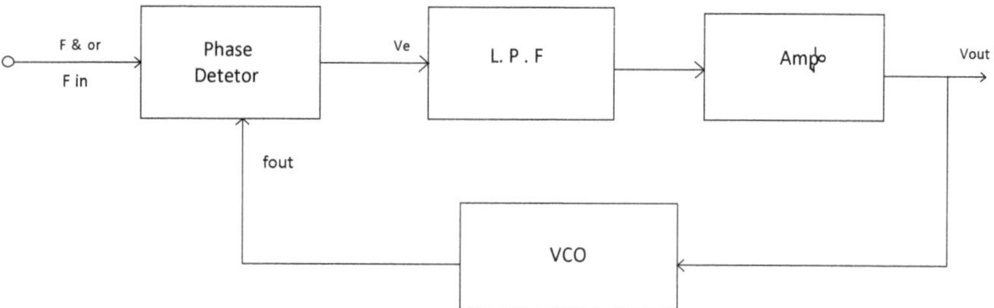

Fig. 6.9 Phase locked loop

"PLL" is available in single package or series of signet ices [SE/NE 5 60}, that is, (560, 561, 562, 564, 565, and 567).

6.5.2.1 Phase Detector Block

Phase locked loop has phase detector comparator which compares the "f_{in}" or "f_s" with the feedback frequency "f_{out}". The output of phase detector is proportional to the phase difference between "f_{in}" and "f_{out}". A phase detector is DC voltage and is referred to as error voltage V_e.

6.5.2.2 Low Pass Filter

The output of phase detector is applied to low pass filter which removes the "high frequency" and produces the DC level, which is applied to the input of amplifier or directly to the VCO.

LPF is also used to establish the dynamic characteristics of the PLL circuit. Filter can be designed by using simple RC network. The dynamic characteristics that are being controlled by filter are known as capture & Lock range. The relationship between the band width & transient response are as filter bandwidth decreases its transient response time increases.

6.5.2.3 Voltage Control Oscillator

It generates input frequency which is directly proportional to the input voltage. Most commonly used VCO IC is SE/NE 566.

PLL works on three possible stages

1. Free running multivibrator.
2. Capture range.
3. Phase look or pull-in time.

(i) **Free running multivibrator**:

When no input signal is applied, then "PLL" is said to be running in free-stage. When input signal is applied, then signal is compared by phase detector and PLL is said to be "in capture range". When output of phase detector is zero, i.e., error signal is zero, by phase detector, then PLL is said to be phase lock with incoming signal.

Lock in range: Once PLL is locked, it can trace the frequency changes in the incoming signal range of frequency, in which PLL can maintain lock with incoming signal is known as "lock in" or "trace range". Usually, it is expressed as a "% of output frequency", i.e., frequency of VCO.

$$\text{frequency}_{\text{lock in time/range}} > \text{frequency}_{\text{capture range}}.$$

(ii) **Capture Range**:

The range of frequency over which the PLL can acquire lock with an input signal is called “capture range”. It is also expressed as a “% of output frequency”.

(iii) **Phase Look or Pull-in time**:

Total time taken by “PLL” to established lock is called “pull-in time”. It depends on, initially, phase and frequency difference between two signals as well as on the overall gain and loop filter characteristics.

6.6 Filter

6.6.1 Simple Active Filters (LP, HP, BP, and Notch Type)

Filters are those circuits which are used to block or attenuate certain band of frequencies or may pass certain band of frequencies. Filters are frequency selective network.

6.6.1.1 Active Filters and Passive Filters

“Active filters” are those which make use of active elements such as “(BJT of OP-AMP)” in their circuit designed.

“Passive filters” are those which make use of passive elements in their circuit passive elements are (RLC).

First-Order Low Pass Filter (Butterworth Filter)

First-order “low pass butterworth filter” uses a “RC network” for filtering OP-AMP which is used in non-inverting configuration as shown in Fig. 6.10. Hence, it does not load down the RC network resistor (R_1) and R_f. Determine the gain of the filter.

$$\frac{V_{out}}{V_1} = \left[1 + \frac{R_f}{R_1}\right] \tag{6.8}$$

$$V_1 = \frac{V_{in}[-jX_C]}{(R - jX_C)} = \frac{V_{in[-j/wc]}}{(R - j/wc)} = \frac{jv_{in}}{(R_{wc} - j)} \tag{6.9}$$

$$V_1 = \frac{+V_{in}}{(R_{wc}j + 1)} \tag{6.10}$$

Since OP-AMP is working in “non-inverting mode”, their output voltage will be

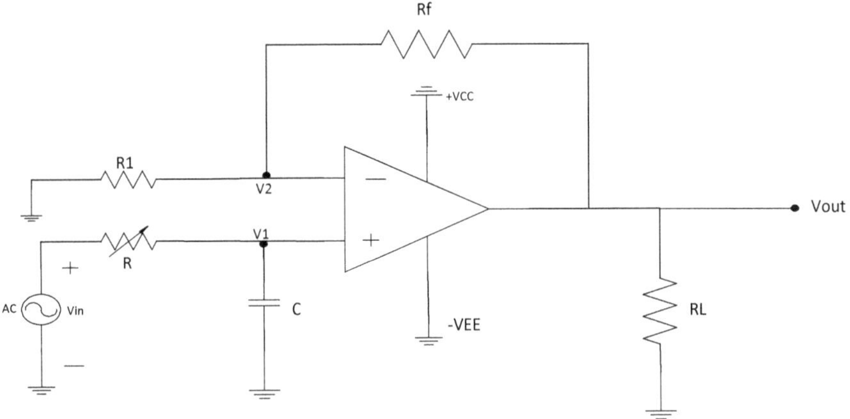

Fig. 6.10 First-order low pass filter

$$V_0 = \left[1 + \frac{R_f}{R_1}\right] V_1. \tag{6.11}$$

Placing the value of "V_1" by using Eqs. (6.8) and (6.10), we get

$$V_{out} = \left[1 + \frac{R_f}{R_1}\right]\left[\frac{V_{in}}{1 + jR_{wc}}\right] \tag{6.12}$$

As

$$(w = 2\pi f) \tag{6.13}$$

$$\frac{V_{out}}{V_{in}} = \frac{A_f}{1 + j2\pi fRC} \tag{6.14}$$

If

$$\left(f_H = \frac{1}{2\pi RC}\right) \tag{6.15}$$

&

$$A_f = \left(1 + \frac{R_f}{R_1}\right) \tag{6.16}$$

Using Eqs. (6.14) and (6.15), now

$$\frac{V_{out}}{V_{in}} = \frac{A_f}{1 + j\left(\frac{f}{f_H}\right)} \tag{6.17}$$

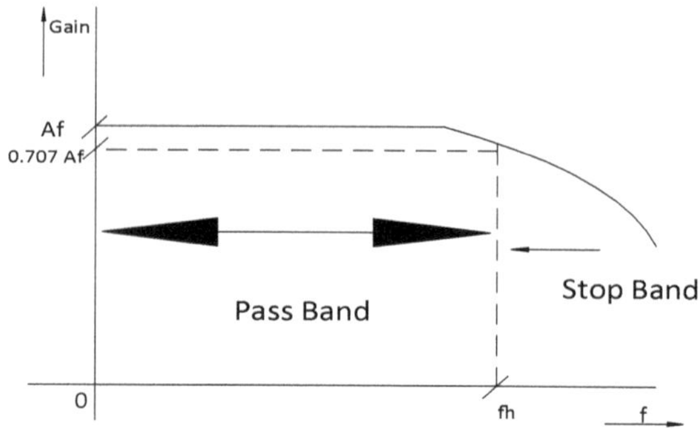

Fig. 6.11 Frequency response curve of first-order low pass filter

where,

f represents input frequency.
A_f is maximum gain.
f_H is higher cut-off frequency.

Magnitude of the gain is given by

$$A_d = \frac{A_f}{1 + (f/fm)^2} = \frac{|V_{out}|}{|V_{in}|}. \tag{6.18}$$

And, phase angle is given by

$$\theta = \tan^{-1}\left(\frac{f}{f_H}\right). \tag{6.19}$$

Figure 6.11 shows the frequency response curve of the first-order low pass filter.

Question 6.1: Design a "LPF" at a cut-off frequency ($f_H = 1$ kHz) and pass band gain = 2.
Solution: Let $C = 0.1$ µf,

$$f_H = 1 \times 10^3\ \text{Hz}.$$

$$f_H = \frac{1}{2\pi RC} \Rightarrow \frac{1}{2\pi * R * 10^{-6}} \Rightarrow 1000\ \text{Hz}.$$

$$R = 1.59\,\text{k}\Omega.$$

$$2 = A_f = 1 + \frac{R_f}{R_1}.$$

$$R_f = R_1 = 10\,\text{k}\Omega.$$

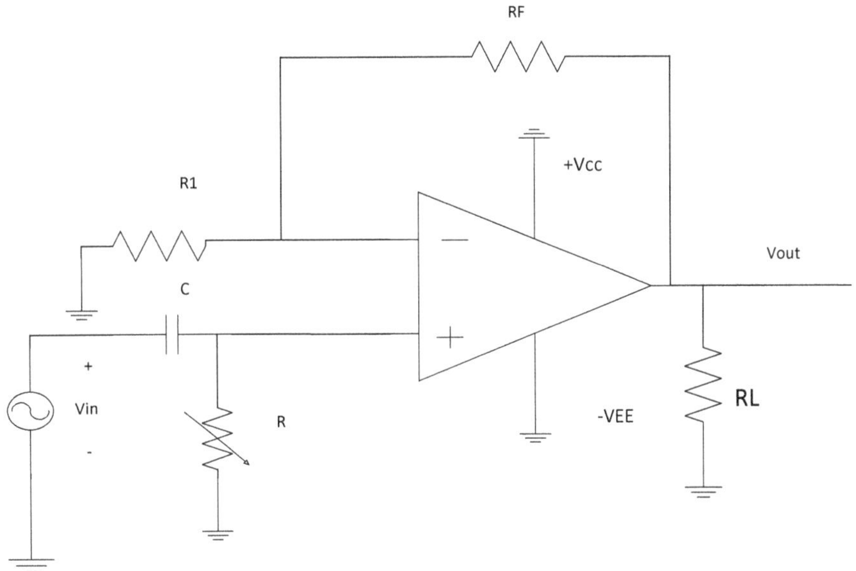

Fig. 6.12 First-order high pass filter

First-Order High Pass Filter ("Butterworth Filter")

HPF is formal (by interchanging frequency determining "R" and "C") in LPF. HPF passes all the frequency greater than lower "cut-off frequency" (f_L), "f_L" is the frequency at which magnitude of the gain is 0.707 times the value of the gain in pass band. Figure 6.12 shows the circuit diagram of high pass filter.

All the frequencies [$f > f_L$] are "pass band frequency" with the highest frequency determine by close loop bandwidth of the "OP-AMP".

$$\text{Gain} = \frac{V_{\text{out}}}{V_{\text{in}}} \tag{6.20}$$

$$V_{\text{out}} = \left[1 + \frac{R_f}{R_1}\right] V_1 \Rightarrow \left[1 + \frac{R_f}{R_1}\right]\left[\frac{R_f}{R + 1/jwc}\right] V_{\text{in}} \tag{6.21}$$

$$\frac{V_{\text{out}}}{V_{\text{in}}} = \text{Gain} = A_d = \left[1 + \frac{R_f}{R_1}\right]\left[\frac{jwcR}{1 + jwRC}\right] \tag{6.22}$$

$$A_d = \left[1 + \frac{R_f}{R_1}\right]\left[\frac{(1 - jwRC)(Rwcj)}{1 + R^2W^2C^2}\right] \tag{6.23}$$

$$A_d = \left(1 + \frac{R_f}{R_1}\right)\left[\frac{R^2W^2C^2 + jRwc}{1 + R^2W^2C^2}\right] \tag{6.24}$$

Figure 6.13 shows the frequency response curve of the first-order high pass filter.

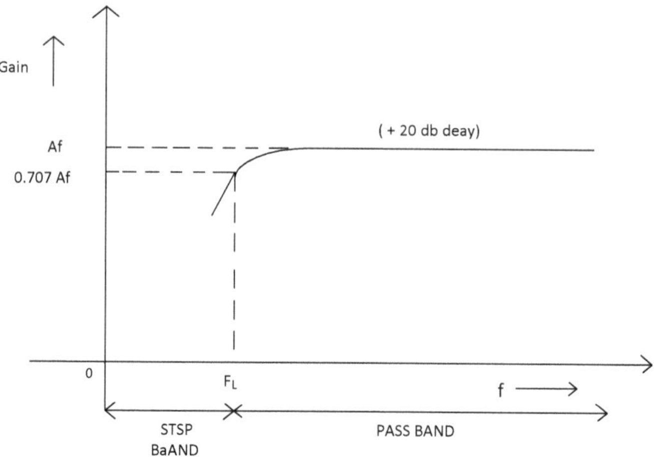

Fig. 6.13 Frequency response of first-order high pass filter

Placing $w = 2\pi f$.

$$A_{\mathrm{d}} = \left(1 + \frac{R_{\mathrm{f}}}{R_1}\right) \frac{\sqrt{\left\{R^2 (2\pi f)^2 C^2\right\} + \{RC 2\pi f\}^2}}{1 + R^2 C^2 (2\pi f)^2} \tag{6.25}$$

$$\& \quad A_{\mathrm{d}} = \left(1 + \frac{R_{\mathrm{f}}}{R_1}\right) \left[\frac{2\pi f RC}{\sqrt{1 + (2\pi f RC)^2}}\right] \tag{6.26}$$

$$A_{\mathrm{d}} = A_{\mathrm{f}} \frac{(2\pi f RC)}{\sqrt{1 + (2\pi f C)^2}}. \tag{6.27}$$

$$\text{If, } f_{\mathrm{L}} = \frac{1}{2\pi \ f RC} = \text{lower cut-off frequency.}$$

Then,
Case 1: $f_{\mathrm{L}}/f > 1, \& \ f_{\mathrm{L}} > f$

$$A_{\mathrm{d}} = \frac{A_{\mathrm{f}}}{\sqrt{1 + \left(\frac{1}{2\pi f RC}\right)^2}} \tag{6.28}$$

where,

$$A_{\mathrm{d}} < A_{\mathrm{f}}.$$

Case 2: $f_{\mathrm{L}}/f < 1, \& \ f_{\mathrm{L}} < f$

$$A_{\mathrm{d}} = \frac{A_{\mathrm{f}}}{\sqrt{1 + \left(\frac{f_{\mathrm{L}}}{f}\right)^2}} \tag{6.29}$$

where,

$$A_{\mathrm{d}} > A_{\mathrm{f}}.$$

Case 3: $f_{\mathrm{L}}/f = 1, \ \& \ f_{\mathrm{L}} > f$

$$A_{\mathrm{d}} = \frac{A_{\mathrm{f}}(f/f_{\mathrm{L}})}{\sqrt{1 + (f/f_{\mathrm{L}})^2}} \tag{6.30}$$

where,

$$A_{\mathrm{d}} = \frac{A_{\mathrm{f}}}{\sqrt{2}}. \tag{6.31}$$

Question 6.2: Design a high pass filter at a "cut-off frequency" of 1 kHz with a pass band gain of "2".

Solution:

$$f_{\mathrm{L}} = \frac{1}{2\pi RC}, \quad \& \ C = 0.01 \times 10^{-6}.$$
$$1 \times 10^3 = \frac{1}{2\pi \times 0.01 \times 10^{-6} R}$$
$$R = 15.9 \ \mathrm{k}\Omega.$$
$$A_{\mathrm{f}} = \left(1 + \frac{R_{\mathrm{f}}}{R_1}\right) \Rightarrow 2 = 1 + \frac{R_{\mathrm{f}}}{R_1} \Rightarrow R_{\mathrm{f}} = R_1$$

Let, $R_{\mathrm{f}} = R_1 = 10 \ \mathrm{k}\Omega$.

Band Pass Filter

It passes frequency between "cut-off frequencies", i.e., (f_{H} & f_{L}) where ($f_{\mathrm{H}} > f_{\mathrm{L}}$), any frequency outside this band is attenuated.

There are two types of BPF given as

1. Wide band pass filter.
2. Narrow band pass filter.

They are distinguished by figure of merits, i.e., "Q". For wide band pass filter, quality factor ($Q < 10$), while for narrow BPF ($Q > 10$). Thus, "Q" is measure of selectivity, i.e., higher the value of "Q" more selectivity will be the filter and its bandwidth will be narrow.

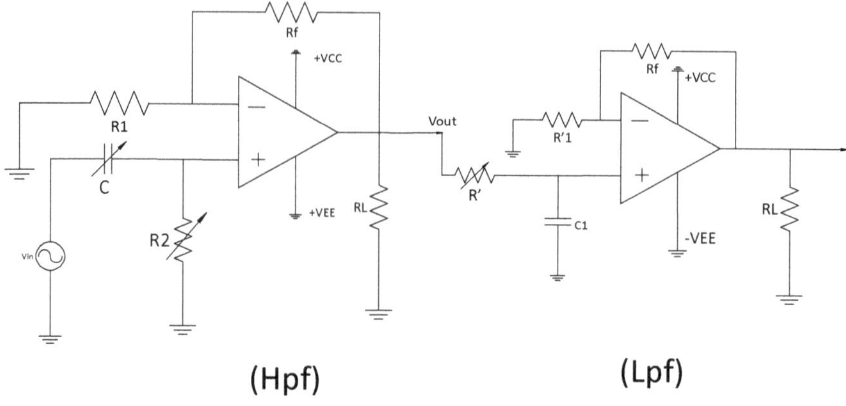

Fig. 6.14 Wide band pass filter

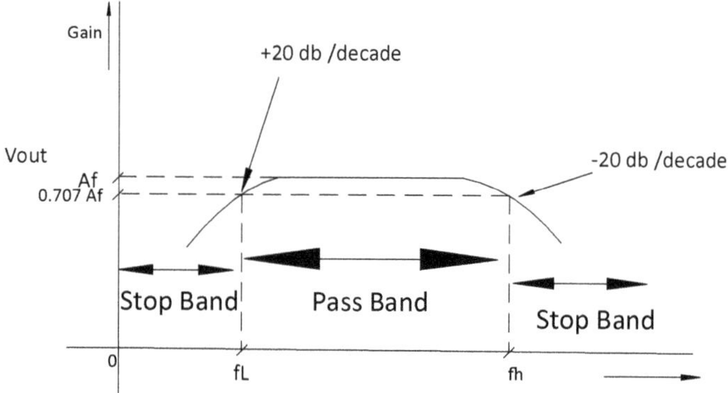

Fig. 6.15 Frequency response curve of wide band pass filter

- **Relationship between "Bandwidth" and "Q" and "f_C"**

$$Q = \frac{f_{\text{Center}}}{\text{BW}} = \frac{\sqrt{(f_H * f_L)}}{(f_H - f_L)} \tag{6.32}$$

where, (f_C = center frequency) $= \sqrt{(f_H * f_L)}$.

Wide Band Pass Filter

This can be formed by simply cascading a high pass filter and low pass section. These high pass and low pass sections can be of first order in which gain is ±20 dB/decay or second order in which gain is ±40 dB/decay. In the given Fig. 6.14, first-order high pass filter connected with first-order low pass filter.

Figure 6.15 shows the frequency response curve of the wide band pass filter.

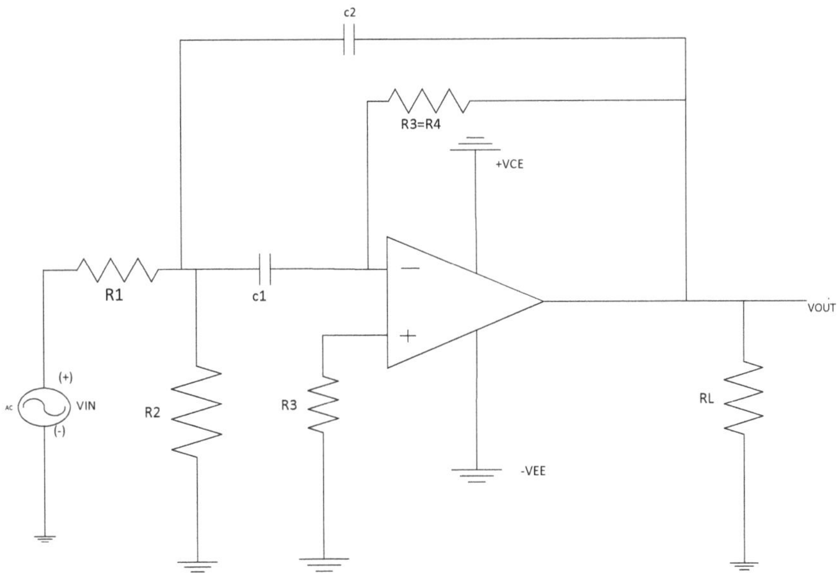

Fig. 6.16 Narrow band pass filter

Question 6.3: Design a "Band Pass Filter" with f_L= 200 Hz, f_H = 1 kHz, and a pass band gain = 4, C_1 = 0.01 × 10^{-6} Pf. Draw the frequency response of the filter. Calculate the value of "Q".

Solution: $A_d = \frac{(1+R_f/R_1)}{\sqrt{[1+(f_L/f)^2]}} \times \frac{(1+R_f/R_1)}{\sqrt{[1+(f/f_H)^2]}}$

1. Low Pass Filter: $f_H = \frac{1}{2\pi R_1 C_1} \Rightarrow R_1 = R_1 = \frac{1}{2\pi C_1 f_H}$, where, $C_1 = 0.01 \times 10^{-6}$ Pf.

$$R_1 = 15.9\ \text{k}\Omega.$$

2. $Q = \frac{\sqrt{f_H \times f_L}}{(f_H - f_L)} = \frac{\sqrt{200.110^3}}{(10^3 - 200)} = 0.56.$
3. High Pass Filter: $f_L = \frac{1}{2\pi R_1 C_1} \Rightarrow R_1 = \frac{1}{2\pi f C_1}$

$$R_1 = 15.9\ \text{k}\Omega.$$

Narrow Band Pass Filter

This filter makes use of multiple feedbacks for narrowing its bandwidth. This narrow band pass filter is unique from the other filters as shown in Fig. 6.16 and consists of:

1. It has two-feedback path, hence it is named as "multiple feedback filter".
2. The "OP-AMP" is used in "Inverting Mode".

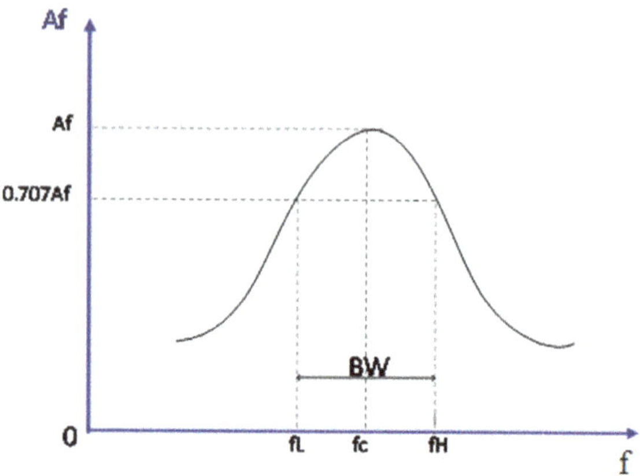

Fig. 6.17 Frequency response curve of narrow band pass filter

Narrow band pass filter is designed for very specific values of center frequency "(f_C)" and "Q", and bandwidth of this filter is very specific.

Value of ($C_1 = C_2 = C < 1$ μf) or can be taken as (0.01 μF). If "Q" is the quality factor, "f_C" is centre frequency, and "A_f" is the gain. The value of

$$\frac{R_3}{2A_f} = R_1 = \frac{Q}{2\pi f_C C A_f}\left[A_f = \frac{R_3}{2R_1}\right] \tag{6.33}$$

$$\frac{R_3}{2\left[2Q^2 - A_f\right]} = R_2 = \frac{Q}{2\pi f_C\left[2Q^2 - A_f\right]}\left(A_f < 2Q^2\right) \tag{6.34}$$

$$R_3 = \frac{Q}{\pi f_C C}. \tag{6.35}$$

Narrow band pass filter resistance (R_1) is connected with supply voltage (V_{in}) for higher values of frequency. In Narrow-Band Pass Filter, resistance (R_1) is connected with supply voltage (V_{in}) for higher values of frequency. For low value of frequency "V_{in}" can be directly connected with the inverting terminal of the OP-AMP in the narrow-band filter.

Figure 6.17 shows the frequency response curve of the narrow band pass filter.

Question 6.4: Design a narrow band pass filter and show that $f_C = 1$ kHz, $Q = 3$, and $A_f = 10$.

Solution:

1. $R_1 = \dfrac{3}{2\pi \times 10^3 \times 0.01 \times 10^{-6} \times 10}$

   $R_1 = 4.77\ \text{k}\Omega.$
2. $R_2 = \frac{3}{2\times3.14\times10^3\times0.01\times10^{-6}(2\times9-10)} \Rightarrow 5.97\ \text{k}\Omega.$
3. $R_3 = \frac{3}{3.14\times10^3\times0.01\times10^{-6}} => 95.5\ \text{k}\Omega.$

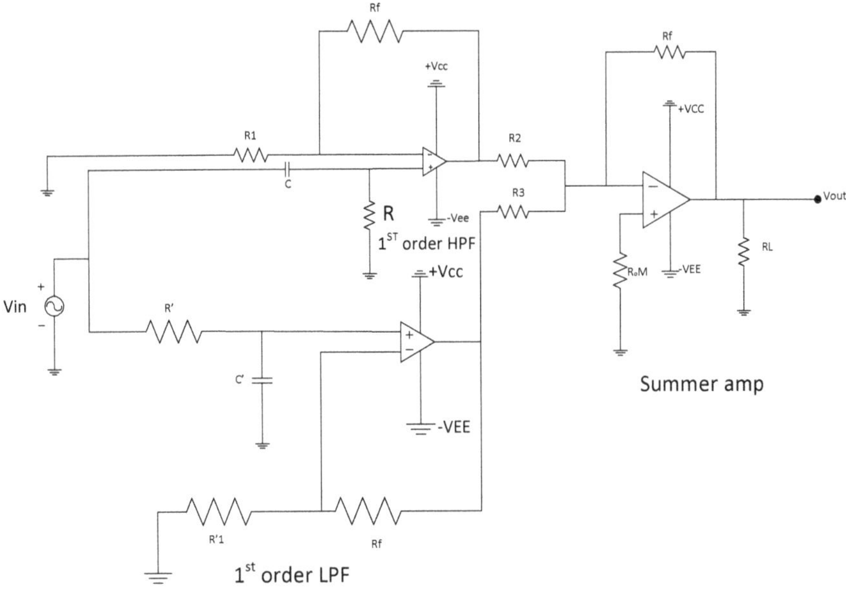

Fig. 6.18 Wide band reject filter

Band Rejection/Band Stop/Band Elimination Filter

Filter frequencies are attenuated in the stop band while they are passed outside the band.

There are two types.

1. Wide band rejection filter.
2. Narrow band rejection filter or notch filter.

Wide Band Rejection Filter

1. It makes use of low pass filter, high pass filter, and summing amplifier as shown in Fig. 6.18.
2. In wide band rejection, lower cut-off frequency is of high pass filter, i.e., (f_L) must be larger than the high cut-off frequency "(f_H)" of LPF.
3. The "pass band gain" of both high pass and low pass section must be equal.
4. Centre frequency "(f_C)" is given by $f_C = \sqrt{(f_H \times f_L)}$.
5. The output frequency of two first-order low pass and high pass is applied through the inverting terminal of summing amplifier "A_3" whose gain is "unity".

Figure 6.19 shows the frequency response curve of the wide band reject filter.

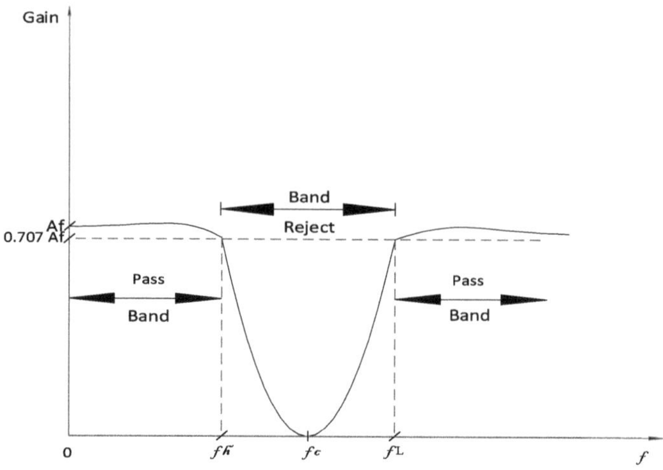

Fig. 6.19 Frequency response curve of wide band reject filter

Question 6.5: Design a wideband rejection filter $f_H = 200$ Hz, $f_L = 1$ kHz.
Solution:

1. Low pass: $f_H = \frac{1}{2RC}$, Let $C = 0.05 \times 10^{-6}$ F.

$$R = \frac{1}{200 \times 2 \times 3.14 \times 0.05 \times 10^{-6}}$$
$$R' = R = 15.9 \text{ k}\Omega.$$

2. High pass: $f_L = \frac{1}{2\pi RC} \Rightarrow R = 15.9\,\text{k}\Omega$. $(C = 0.01 \times 10^{-6}\,\text{F})$

$$A_f = \left(1 + \frac{R_f}{R_f}\right) \Rightarrow 2 = \left(1 + \frac{R_f}{R_f}\right) \Rightarrow (R_f = R_1).$$
$$R_1 = R_f = 10 \text{ k}\Omega = R_1^{'} = R_f^{'}.$$

Narrow Band Rejection Filter or Notch Filter

1. In narrow band pass filter, the quality of factor (Q) is greater than 10, and band width is much narrow as compare to the wide Band Pass Filter.
2. Narrow band rejection filter is also known as "notch filter".
3. It is used for rejection of signal frequency such as 60 kHz.
4. Most commonly used notch filter is "Twin–T network" which consists of passive filter composed of "Two-Twin-T-network".

$$f_N = \frac{1}{2\pi RC} \tag{6.36}$$

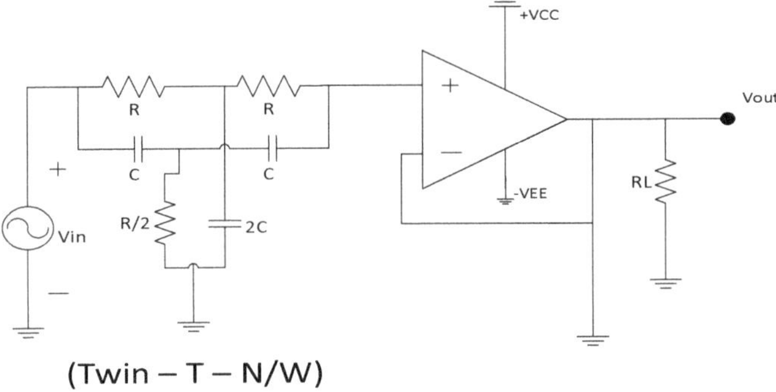

Fig. 6.20 Narrow band reject filter

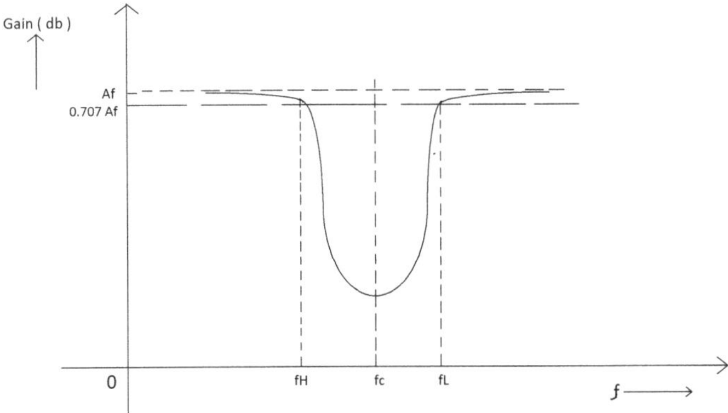

Fig. 6.21 Frequency response curve of narrow band reject filter

First T network consists of "2-R" and "1-C", and second T network consists of "2-C" and "1-R", which are connected through the non-inverting terminal of OP-AMP as shown in Fig. 6.20.

Narrow BRF is used to reject the single or specific frequency from a given band of frequencies.

Figure 6.21 shows the frequency response curve of the narrow band rejection filter.

All Pass Filter

It passes all the frequency components of the input signal without "attenuation" while providing predictable "phase-shift" for different frequencies of the input signal, when signals are transmitted over transmission lines such as "telephone wire". The under frequencies change in-phase, in order to compensate to this phase change, all pass

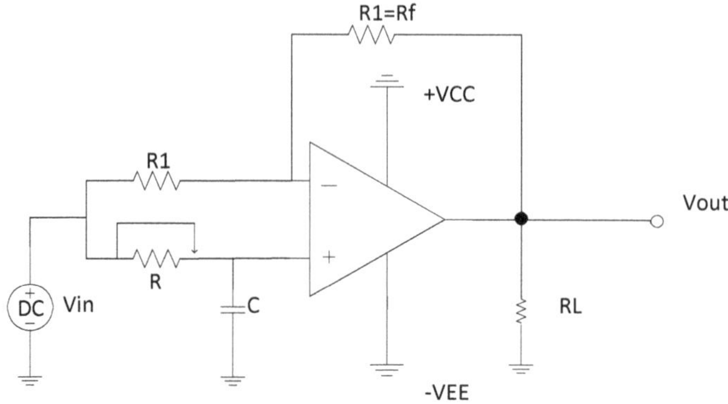

Fig. 6.22 All pass filter

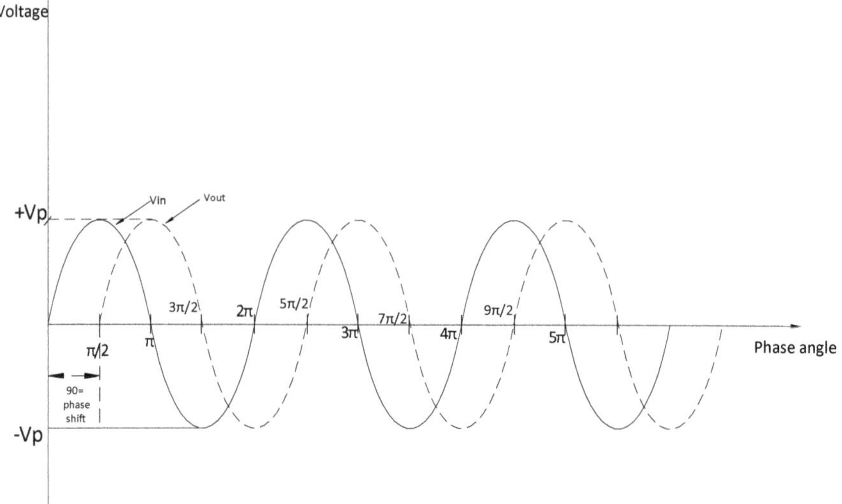

Fig. 6.23 Frequency response curve of all pass filter

filters are used. All pass filters make use of "superimposition" of two signals applied at "inverting" and "non-inverting" terminals of "OP-AMP" as shown in Fig. 6.22.

Figure 6.23 shows the frequency response curve of the all pass filter.

Gain of all pass filter is

1. $$A_{\mathrm{d}} = \frac{V_{\mathrm{out}}}{V_{\mathrm{in}}} = \left(\frac{1 - j2\pi f RC}{1 + j2\pi f RC}\right). \tag{6.37}$$

The phase of all pass filter is

2. $$\phi = -2 \times \tan^{-1}(w \times R \times C). \tag{6.38}$$

6.7 IC 555 Timer and Its Applications

In Fig. 6.24, the integrated circuit 5 KΩ-resistor connected in series. IC 555 is a monolithic which is used as a “delay circuit”. It was first developed by Signatic Corporation as “SE/NE-555”. It can be used in mono-stable and A-stable multivibrator, digital logic, proper waveform generator, analog frequency meter, tachometer, and AC-DC converter, etc.

1. It is used to produce time delay or oscillations. It is available in metal tank.
2. It has 14 Pin “DIP” (dual input package).
3. “SE-555” is defined for temperature range, i.e., (−55 to +125 °C), and any 555 operates over a voltage range of (+5 to +18 V).
4. Its duty cycle is adjustable from μsec to hours. It can source or sink current up to 200 mA.

6.7.1 Application of IC 555

1. Used in timer circuit.
2. Used in pulse generator,
3. Used in oscillator, and frequency divider.
4. Used in PWM, missing pulse detection, capacitance measurement, and so on.

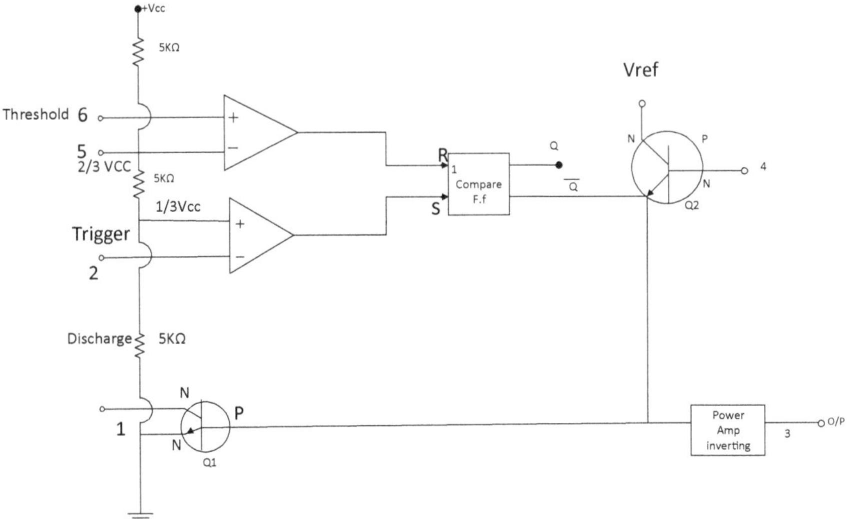

Fig. 6.24 IC 555 timer

6.8 Voltage Regulator Introduction

Voltage regulator is used to provide stable DC voltage of power in electronic circuit. Voltage regulator should be capable of providing large output current.

6.8.1 *Types of Voltage Regulator*

There are three types of voltage regulators named as

1. Series regulator.
2. Shunt regulator.
3. Switching regulator.

6.8.1.1 Series Regulator

It makes use of power transistor connected in series with unregulated DC input and the load. The output voltage is controlled continuous voltage drop taking place across series power transistor.

Since the transistor conducts in "active region" or "linear region", those regulations are also called "linear regulation". Linear regulation may have fixed or variable output and can be positive or negative. They are provided with series of 78XX, 79XX, and 723-ICC.

"Series regulator" is an electronic circuit that provides stable DC voltage independent of load current, temperature, AC line variations. It consists of four parts

(i) Reference voltage circuit.
(ii) Error amplifier.
(iii) Series power transistor.
(iv) Feedback network.

1. In the above Fig. 6.25, a power transistor Q_1 is in series with unregulated DC voltage input voltage (V_{in}) and regulated output (V_{out}). So, it must be observed the difference between their two voltages, whenever any fluctuation in V_{out} occurs.
2. The transistor Q_1 is also connected as voltage follower. Therefore, its output is proportional to the input. It also provides self-current gain to drive the load.
3. The fluctuation of output is feedback to the inverting terminal of error amplifier OP-AMP.
4. Non-inverting terminal of error amplifier is connected with zener diode of constant voltage providing "V_{Ref}".
5. The sampled voltage is compared to the V_{Ref}, the output of error amplifier V_0' drive power transistor Q.

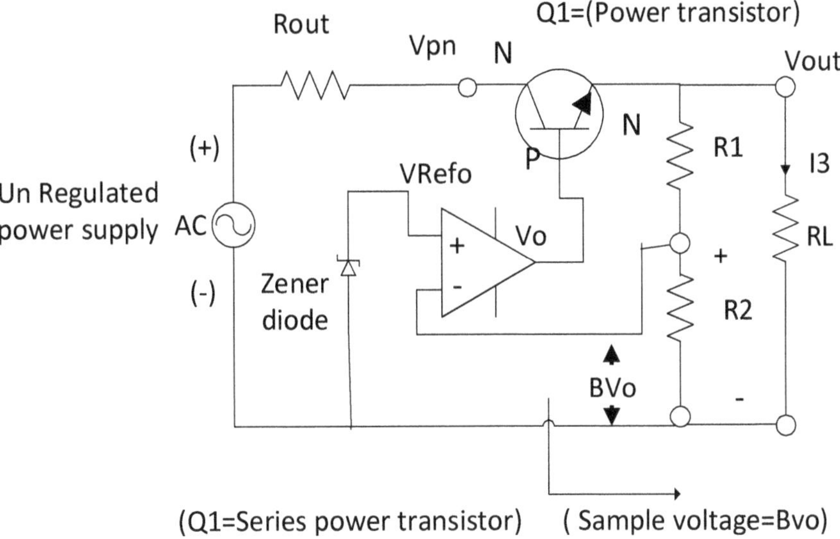

Fig. 6.25 Series regulator

6. The V_{out} increases due to variation in load current and then sample voltage also increases which is given by βV_0 where, $\beta = R_2/(R_1 + R_2)$.
7. If feedback voltage is greater than $V_{Ref,}$ then the V_{out} is 180° out of phase and also decreases.
8. Hence, the increment in V_{out} is nullified or reduced. Similarly, output reduction also gets regulated. Therefore, the V_{out} is regulated by voltage regulator.

6.8.1.2 Shunt Regulator

The schematic, shown in Fig. 6.26, is that of a shunt-voltage regulator. Notice that Q_1 is in parallel with the load. Components of this circuit are identical with these of the series voltage regulator except for the addition of fixed R_S. As you study the schematic, you will see that this R is connected in series with the output load R. The current limiting resistor R_1 and zener diode (CR1) provide a constant reference voltage for the base-collector junction of Q_1. Notice that the bias of Q_1 is determined by the voltage drop across R_S and R_1. As you should know, the amount of forward bias across a transistor affects its total resistance. In this case, the voltage drop across R_S is the key to the total circuit operation.

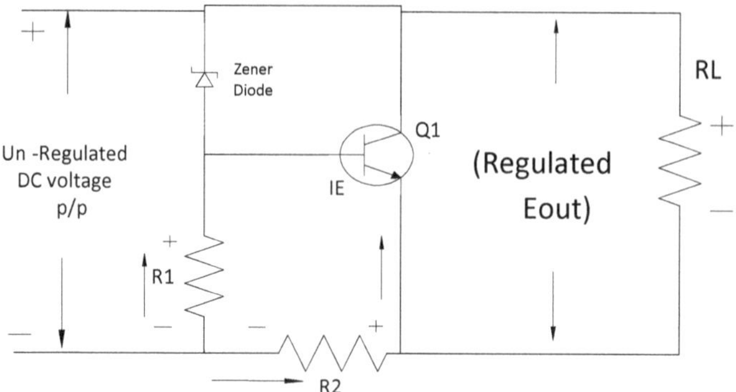

Fig. 6.26 Shunt regulator

6.8.1.3 Switching Regulator

It is used to provide output in switching mode. Its overall efficiency is higher as compared to fixed regulator or adjustable regulator. Since, it operates for the finite time period interval depending upon pulse width applied to switch.

$$V_{out} = \frac{t_{ON}}{T} \times V_{in} \tag{6.39}$$

where,

T shows total time (ON-OFF time).
t_{ON} is ON time.

Let assuming that if V_{in} and T both are constant, then

$$V_{out} \propto t_{ON} \tag{6.40}$$

In Fig. 6.27, when the input voltage V_{in} applied to the switch, i.e., (BJT or FET), and this switch is attached with pulse regulator by which we can control the switch, i.e., ON and OFF time. If the width of pulse regulator is high, then switch takes more time, otherwise vice versa.

Further, this AC signal goes on the filter, i.e., (combination of R-L-C), and capacitor blocks DC signal and passes the AC signal. Finally, we get the AC supply at the output terminal V_{out}.

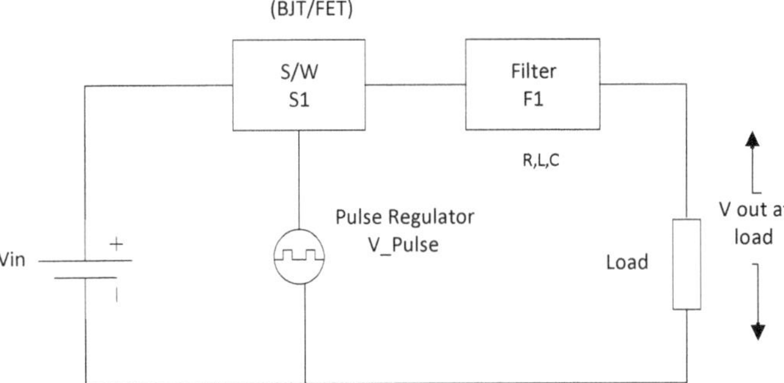

Fig. 6.27 Switching regulator

6.9 OP-AMP-Based IC Regulator

6.9.1 Fixed Voltage Regulator

The fixed voltage regulator has an unregulated DC input voltage V_{i}, applied to one input terminal, a regulated output DC voltage V_0 from a second terminal and third terminal connected to ground.

6.9.1.1 Fixed Positive Voltage Regulator

In Fig. 6.28, the series 78 regulators provide fixed regulators voltage from 5 to 24 V. An unregulated input voltage V_{i} is filtered by capacitor C_1 and connected IC's IN terminal. The ICs OUT terminal provides a regulated +12 V, which is flittered by capacitor C_2. The third IC terminal is connected to ground, whereas the input voltage may vary over some permissible voltage range, and the output load may vary over some acceptable range, and the output voltage remains constant within specified voltage variation limits.

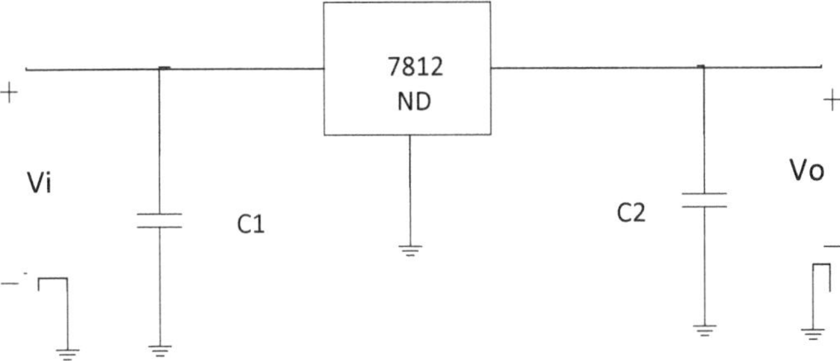

Fig. 6.28 Positive fixed voltage IC regulator

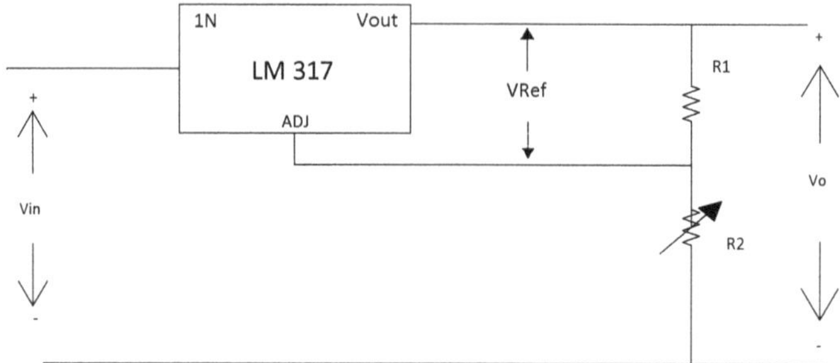

Fig. 6.29 Adjustable voltage IC regulator

6.9.1.2 Fixed Negative Voltage Regulator

The series 7900 ICs provide negative regulation, similarly to those providing positive voltages.

6.9.2 Adjustable Voltage Regulator

Voltage regulators are also provided in circuit configuration that allows the users to set the output voltage regulated value. In Fig. 6.29, resistors R_1 and R_2 set the output at any desired voltage over the adjustment range (1.2–37 V).

The output voltage desired can be regulated using

$$\left[V_0 = V_{\text{ref}}\left(1 + \frac{R_2}{R_1}\right) + I_{\text{adj}} R_2\right] \tag{6.41}$$

with IC values of $V_{\text{ref}} = 1.25\,\text{V}$, & $I_{\text{adj}} = 100\,\mu\text{A}$.

6.10 Sample and Hold Circuit

In sample and hold circuit, firstly, we sample the input signals & hold-on it's last sample value until the previous input signal get sampled. Sampler used in sample and hold circuit is "E-type MOSFET".

Working of sample and hold circuit:

(i) In the Fig. 6.30, the output voltage of E-MOSFET is controlled by sample & hold voltage V_S, and the capacitor is used as storage elements.

(ii) When the input analog signal, which is to be sampled, is applied to the E-MOSFET, whose gate terminal is supplied with S/H voltage during positive portion of V_S, E-MOSFET conducts and acts as a close switch.

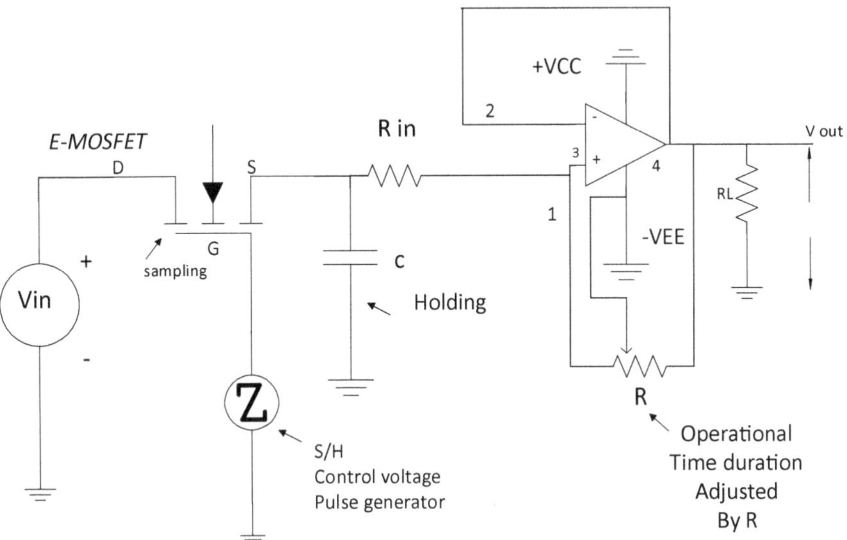

Fig. 6.30 Sample and hold circuit

(iii) V_{in} has charge the capacitor "*C*", i.e., V_{in} appears across *C* and provides the output.

(iv) When $V_S = 0$, or negative E-MOSFET is OFF, i.e., non-conducting, and acts as an open switch. The only discharge path for capacitor "*C*" is through OP-AMP.

(v) Since, the R_{in} of OP-AMP voltage follower is very high. The time period of sample and controlled voltage V_S during which the voltage across the capacitor "*C*" is equal to input voltage, i.e., $V_C = V_{in}$, is called sample period.

(vi) The t_H (time period) of V_S during which the voltage across the *C* is constant are called hold period.

(vii) The output of OP-AMP is observed during voltage period, to obtain output of the input waveform. The frequency of sample and hold circuit (f_{sh}) control voltage must be higher than that of the input (f_{in}). ($f_{sh} > f_{in}$).

(viii) The low-leakage capacitance such as "telephone" is generally chosen in some applications.

(ix) Precision or high-speed OP-AMP is used.

Figure 6.31 depicts the output waveform of sample and hold circuit. "V_p" is the input signal waveform, "V_{in}" is the clock pulse generated by E-MOSFET. Finally, the output of sample and hold circuit is denoted as by "V_{out}" which is the combination of V_p and V_{in}.

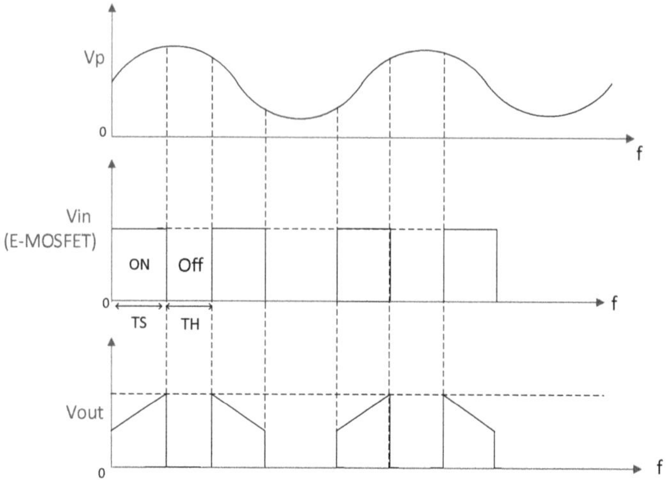

Fig. 6.31 Output waveform of sample and hold circuit

6.11 Analog-to-Digital "A/D" Converter

This method uses ladder network along with counter and comparator circuits as shown in Fig. 6.32. A digital counter advances from a zero count while a ladder network driven by counter output a staircase voltage, which increases one voltage increment for each count step. A comparator circuit, receiving both staircase voltage

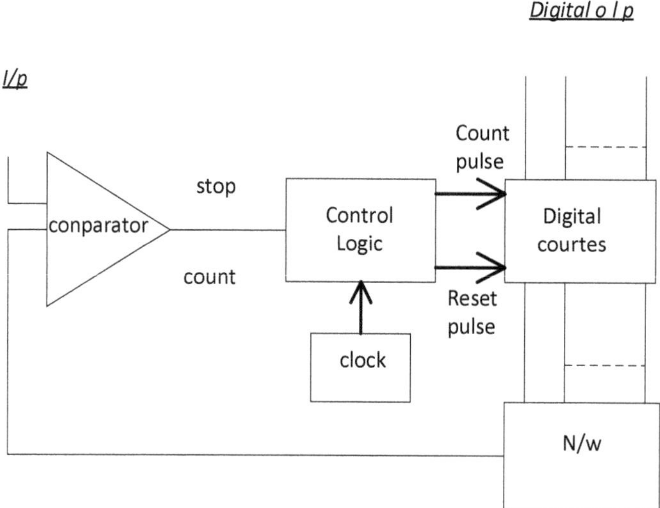

Fig. 6.32 A/D converter using OP-AMP

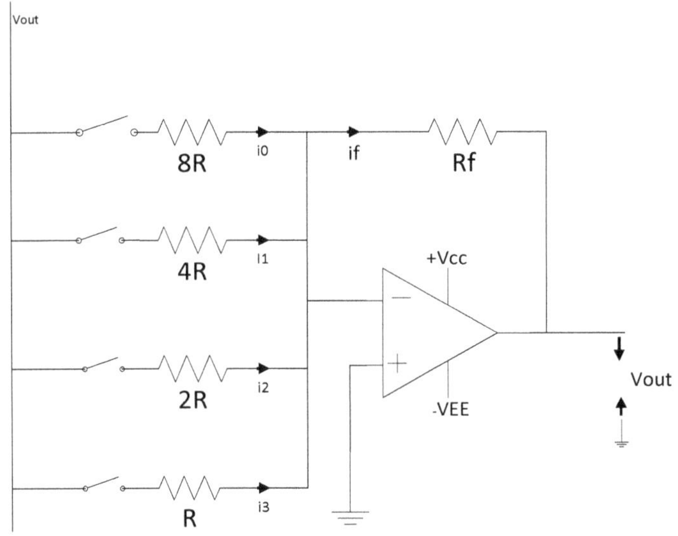

Fig. 6.33 D/A converter using OP-AMP as summing amplifier

and along input voltage, provides a signal to stop the count when the staircase voltage rises above the input voltage. The counter value at that time is the digital output.

The amount of voltage charge stepped by staircase signal depends on the number of count bits used. A converter using fewer count stages would carry conversion per second. The conversion accuracy depends on the accuracy of comparator.

6.12 Digital-to-Analog "D/A" Converter

The basic circuit of an OP-AMP used as an inverter is shown in Fig. 6.33. The input signal is applied to the inverting terminal of OP-AMP. The input impedance is almost infinity and almost no current can enter the input terminals. The input voltage is fixed to almost zero at its input terminals. The upper terminal is at nearly the same pot as the ground terminal. Therefore, the upper terminal is called a virtual ground.

The difference between virtual ground and normal ground is zero while sinking any amount of current. Virtual ground is for voltage, not for current. Therefore, virtual ground point is at 0 V, i.e., it sinks current

As

$$I_{\text{in}} = \frac{V_{\text{in}}}{R_{\text{in}}}, \quad \text{where, } I_{\text{in}} \cong I_{\text{F}} \tag{6.42}$$

Because, no current can enter the virtual ground (because of ∞ impedance), all the current has to go through R_F with respect to virtual ground. V_{out} is measured with respect to normal ground.

Summary

- **This chapter included introduction of multivibrators and their classifications.**
- **Describe A-stable, Mono and Bi-stable multivibrators along with applications.**
- **Comparative study between all multivibrators.**
- **Schmitt trigger and its working.**
- **Introduce phase locked loop and VCO with block diagrams.**
- **Introduce filters like LPF, HPF, BPF, BRF, and all pass filter with their gain and phase angle mathematical expression.**
- **Wide band and narrow band BPF or BSF filter.**
- **IC 555 with PIN configuration and applications.**
- **Discuss voltage regulator and their classification as series, shunt, and switching voltage regulator.**
- **OP-AMP-based voltage regulators, fixed voltage regulator, and adjustable voltage regulator.**
- **Define S/H system with operational working.**
- **A/D and D/A converter using OP-AMP.**

Numerical Problems

Question 1: Write down the differences between A-stable, mono-stable, and bi-stable multivibrators.
Question 2: Draw and explain phase locked loop.
Question 3: Designed a high pass filter (HPF) at a "cut-off frequency" of 10 kHz with a pass band gain of "4".
Question 4: Explain and draw circuit diagram of IC-555.
Question 5: What is switching voltage regulator? Explain it.
Question 6: Designed by "Band Pass Filter" with f_L= 100 Hz, f_H = 20 kHz, and a pass band gain = 4, $C_1 = 0.001 \times 10^{-3}$ nf. Compute the value of "Q".
Question 7: Explain sample and hold circuit with its waveform.
Question 8: Explain first-order low pass filter (LPF), and derive the mathematical expression of gain and phase response for LPF.
Question 9: What is adjustable and fixed voltage regulator?
Question 10: Draw and explain analog-to-digital converter circuitry.
Question 11: Explain voltage control oscillator.

Chapter 7
Multistage and Tuned Amplifiers

Learning Objectives

- **Introduction to multistage amplifiers with applications.**
- **Calculate the gain of amplifiers, and draw frequency response curve.**
- **Introduce coupling and their classifications.**
- **Analyze darling ton amplifier.**
- **Discuss and drive bootstrapping amplifier.**
- **Discuss tuned and double-tuned voltage amplifiers.**

7.1 Introduction to Multistage Amplifiers

In multistage amplifier, we feed the output of stage to the input of the next stage as shown in Fig. 7.1. This arrangement of amplifier is known as "cascade." A multistage amplifier using two or more single-stage common emitter amplifier is known as "cascaded" amplifier. Figure 7.2 shows the *n*-stage cascaded amplifier.

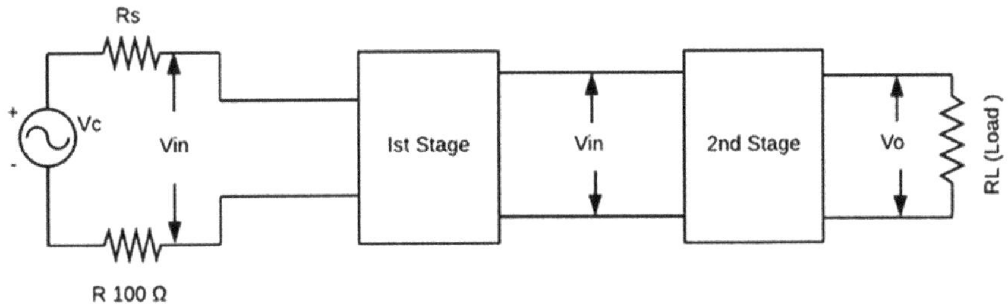

Fig. 7.1 Two-stage cascaded amplifier

G. S. Tomar and A. Bagwari, *Fundamentals of Electronic Devices and Circuits*, Algorithms for Intelligent Systems, https://doi.org/10.1007/978-981-15-0267-5_7

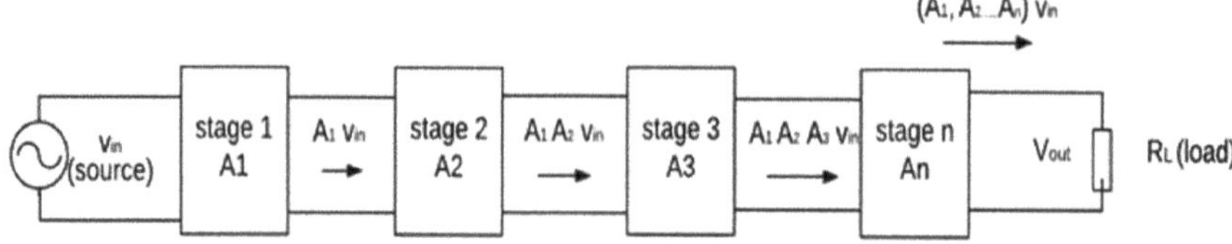

Fig. 7.2 *n*-Stage cascaded amplifier

7.1.1 Applications of Multistage Amplifier

Amplifier used in radio and TV receiver is usually multistage amplifier.

7.1.2 Calculation of Amplifier Gain

Output gain of amplifier first stage

$$\text{Gain} = A_1 = \frac{\text{output}}{\text{input}} = \frac{V_{\text{out}}}{V_{\text{in}}} \Rightarrow V_{\text{out}} = A_1 V_{\text{in}}. \tag{7.1}$$

$$\text{Output of amplifier second stage} = A_1 A_2 V_{\text{in}}.$$

$$\vdots$$

$$\text{Output of amplifier } n\text{th stage} = (A_1 A_2 A_3 A_4 \ldots \mathrm{A}_n) V_{\text{in}}. \tag{7.2}$$

As we know that gain of any system is the ratio of output to input, i.e.,

$$\text{Gain} = \frac{\text{Output}}{\text{Input}} = \frac{V_{\text{o}}}{V_{\text{i}}} \tag{7.3}$$

$$\text{Gain} = \frac{(A_1 A_2 A_3 \ldots A_n) V_{\text{i}}}{V_{\text{i}}} \tag{7.4}$$

Finally, the mathematical expression for amplifier gain is given as

$$\text{Gain} = A_1 A_2 A_3 \ldots A_n \tag{7.5}$$

Further, in terms of decibel, power gain of amplifier can be computed as

$$A_{\text{P}} = \text{Gain}|_{\text{dB}} = \log_{10}\left(\frac{P_{\text{o}}}{P_{\text{i}}}\right) \text{dB} \tag{7.6}$$

- **Special Case**:
  1. Voltage gain of amplifier in terms of dB

$$A_V = \text{Gain}|_{\text{dB}} = 20\log_{10}\left(\frac{V_o}{V_i}\right)\text{dB} \tag{7.7}$$

  2. Current gain of amplifier in terms of dB

$$A_I = \text{Gain}|_{\text{dB}} = 20\log_{10}\left(\frac{I_o}{I_i}\right)\text{dB} \tag{7.8}$$

- **Gain of Multistage amplifier in dB**:
  The gain of multiple stage is

$$\text{Gain} = A_1A_2A_3\ \ldots\ A_n$$

Now, the overall gain of the multistage amplifier will be computed in (dB) as under

$$20\log_{10} A = 20\log_{10} A_1 + 20\log_{10} A_2 + 20\log_{10} A_3 + \cdots + 20\log_{10} A_n \tag{7.9}$$

$$A_{|\text{dB}} = A_{1|\text{dB}} + A_{2|\text{dB}} + A_{3|\text{dB}} + \cdots + A_{n|\text{dB}} \tag{7.10}$$

7.1.3 Frequency Response of an Amplifier

The frequency response of an amplifier is the graph of amplifier gain in dB plotted against the frequency. It shows the variation in amplifier gain with respect to frequency. In other words, frequency response of an amplifier shows the ratio between frequency and gain of the amplifier. The bandwidth of an amplifier is the 3 dB down frequency range of the maximum gain of the amplifier. Hence, the fall in voltage gain form maximum gain will be

$$=> 20\log_{10}\left(\frac{A_V}{0.707 \times A_V}\right) => 3\,\text{db}.$$

7.1.3.1 Ideal Frequency Response

In ideal frequency response, the gain should be flat (constant) over the entire frequency range.

7.1.3.2 Practical Frequency Response

In practical response, the gain of amplifier does not flat over the entire frequency range.

7.1.3.3 Midband Gain

We define the midband of an amplifier as the band of frequencies between $10\,f_1$ and $0.1\,f_2$ as frequency shown in Fig. 7.3. In the midband, the voltage gain of the amplifier is approximately maximum. It is denoted as midband gain or amid gain. Although an amplifier normally operates in the midband, there are times when we want to know what the voltage gain is outside of the midband.

The voltage gain of amplifier outside the midband or gain for midfrequency range approximately is given by

$$A = \frac{\text{output}}{\text{input}} \tag{7.11}$$

$$A = \frac{A_{\text{mid}}}{\left[\sqrt{\left[1 + \left(\frac{f_1}{f_2}\right)^2\right] \times \left[1 + \left(\frac{f}{f_2}\right)^2\right]}\right]}. \tag{7.12}$$

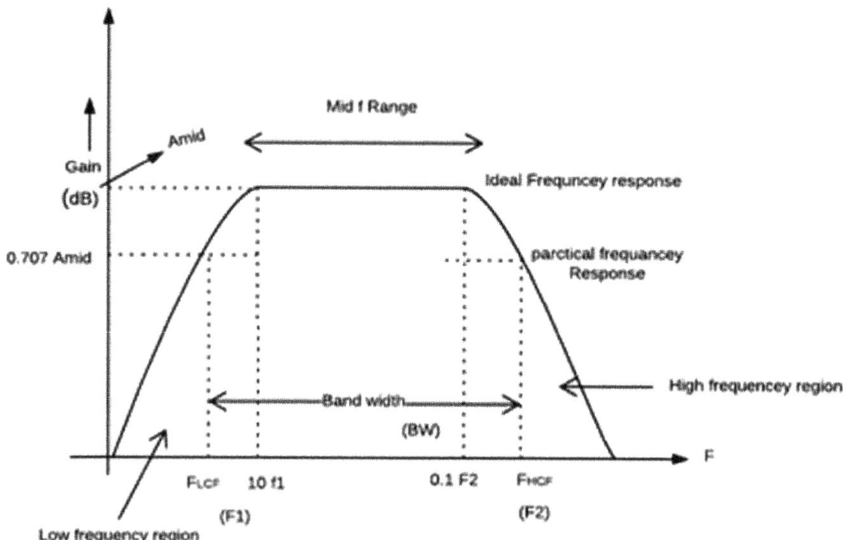

Fig. 7.3 Frequency response of an amplifier

(i) **Midband Gain**: In the midband gain, $\frac{f_1}{f} = 0$, and $\frac{f}{f_2} = 0$, now

$$A = A_{\text{mid}}. \tag{7.13}$$

(ii) **Below Midband Gain**: The condition for low-frequency range is $\frac{f}{f_2} = 0$, and the voltage gain of amplifier for low-frequency range approximately is given by

$$A = \frac{A_{\text{mid}}}{\sqrt{1 + \left(\frac{f_1}{f}\right)^2}}. \tag{7.14}$$

(iii) **Above Midband Gain**: The condition for low-frequency range is $\frac{f_1}{f} = 0$, and the voltage gain of amplifier for high-frequency range approximately is given by

$$A = \frac{A_{\text{mid}}}{\sqrt{1 + \left(\frac{f}{f_2}\right)^2}}. \tag{7.15}$$

Question 7.1 For an amplifier midband Gain = 100 and lower cut-off frequency = 1 kHz, find the gain of amplifier at 20 Hz.

Solution $A_{\text{mid}} = 100,\ f_1(\text{LCF}) = 1\ \text{KHz},\ f = 20\ \text{Hz}.$
Now, the gain of amplifier will be

$$A = \frac{A_{\text{mid}}}{\sqrt{1 + \left(\frac{f_1}{f}\right)^2}}$$
$$A = \frac{100}{\sqrt{1 + \left(\frac{1\times10^3}{20}\right)^2}}$$
$$A = \frac{100}{\sqrt{1 + 2500}}$$
$$A = \frac{100}{50.009999}$$
$$A = 1.9996 \cong 2$$
$$A = 2.$$

Question 7.2 For an amplifier, 3 dB gain is 200 and higher cut-off, $f = 20$ kHz, find gain of amplifier at, $f = 100$ kHz.

Solution Gain at 3 dB = 200 (without dB).

$$f_2 = 20\,\text{kHz}, \quad f = 100\,\text{kHz},$$

$$A_{mid} = \frac{200}{0.707}$$
$$A_{mid} = 282.88.$$

$$A = \frac{A_{\text{mid}}}{\sqrt{1 + \left(\frac{f}{f_2}\right)^2}}$$
$$A = \frac{A_{\text{mid}}}{\sqrt{1 + \left(\frac{100\times 10^3}{20\times 10^3}\right)^2}}$$
$$A = \frac{282.88}{\sqrt{1+25}}$$
$$A = \frac{282.88}{5.0999}$$
$$A = 55.4773.$$

7.2 Cascade Amplifiers

As a matter of fact, a cascade amplifier has one transistor on the top of (in series with) another. This is illustrated in Fig. 7.4, in this figure, transistor "Q_1" and its associated components operate as a common emitter input stage, whereas the circuit of transistor "Q_2" operates as a common base output stage. This type of arrangement is designed to provide a high Z_{in} with low voltage gain to ensure that the input Miller capacitance is at a minimum with the common base stage providing good high-frequency operation.

For DC conditions of this circuit, it may be observed that the I_{E} for transistor Q_1 is set by V_{E1} and R_4. Also, the collector current I_{C1} is approximately equal to I_{E1} and I_{E2} equals I_{C2}. This means that the current I_{C2} is approximately equal to I_{E1}, and this current I_{C2} will remain constant irrespective of the value of voltage V_{B2}, as long as V_{CE1} remains large enough for correct operation of the transistor Q_1.

Further, the Z_{in} to the emitter of transistor Q_2 constitutes the AC load in the collector circuit of transistor Q_1. For stage 1, the voltage gain will be

$$A_{\text{V1}} = \frac{-h_{\text{fe}}}{h_{\text{ie}}} * (Z_{\text{in}} \text{ to } Q_2) \tag{7.16}$$

$$= \frac{-h_{\text{fe}}}{h_{\text{ie}}} * \left[\frac{h_{\text{ie}}}{1 + h_{\text{fe}}}\right] \tag{7.17}$$

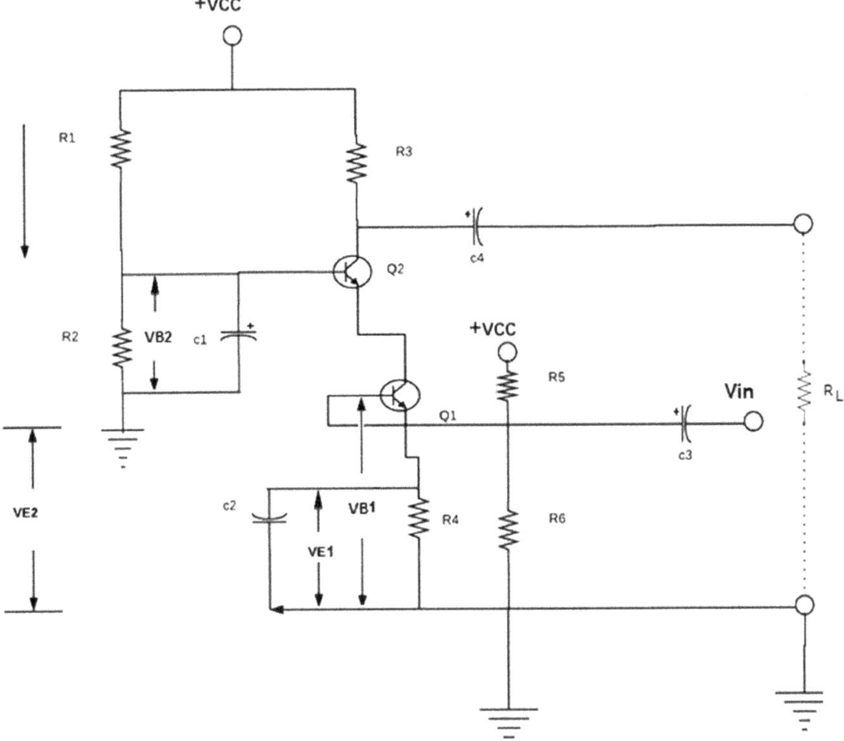

Fig. 7.4 Cascade amplifier

$$\cong -1 \tag{7.18}$$

Thus, with a stage gain of only 1, no Miller effect occurs at transistor Q_1. Now, voltage gain for stage 2 is given by,

$$A_{V2} = h_{fb} * \left[\frac{R_3 || R_L}{h_{ib}} \right] \tag{7.19}$$

Converting to common emitter parameters, the overall voltage gain for the cascade amplifier will be same as that for a common emitter amplifier. Thus, the overall voltage gain

$$A_V = -h_{fe} * \left[\frac{R_3 || R_L}{h_{ie}} \right] \tag{7.20}$$

This means that a cascade amplifier has high Z_{in}, the same voltage gain as a common emitter amplifier but very low input capacitance.

7.3 Coupling of Amplifiers

All amplifiers need some kind of coupling network. The multistage amplifier needs coupling between their individual stages. This type of coupling is known as "inter-stage coupling." It serves two purposes as under

1. It transfers AC output of one stage to the input of the next stage.
2. It isolates the DC conditions of one stage to the next stage.

7.4 Types of Coupling of Amplifiers

There are four types of coupling schemes.

7.4.1 Resistance coupling "RC" coupling.
7.4.2 Impedance coupling.
7.4.3 Transfer coupling.
7.4.4 Direct coupling or DC amplifier coupling.

7.4.1 Resistance Coupling "R-C" Coupling

In this method, the signal develops across the collector resistor of each stage is coupled through the capacitor "*C*" into the base of the next stage.

The cascaded stages amplify the signal, and the overall gain is equal to the product of individual stage gains as shown in Fig. 7.5. The amplifiers using this coupling scheme are known as "R-C coupled amplifiers."

- **Advantages**
  1. By using RC amplifier, we are able to achieve high amplifier gain.
  2. Losses are less.

7.4.2 Impedance Coupling

In this Fig. 7.6, it can be noted that the collector resistance R_C is replaced by inductor *L*, and as frequency increases, the inductive reactance X_L also increase. Due to this, inductor passes DC current and blocks alternate current (AC). The amplifier is using this type of scheme known as impedance coupling.

In the given figure, it can be noted that the collector resistor (R_C) is replaced by an inductor (L_1). As the frequency increases, the inductive reactance X_L ($\approx wL$)

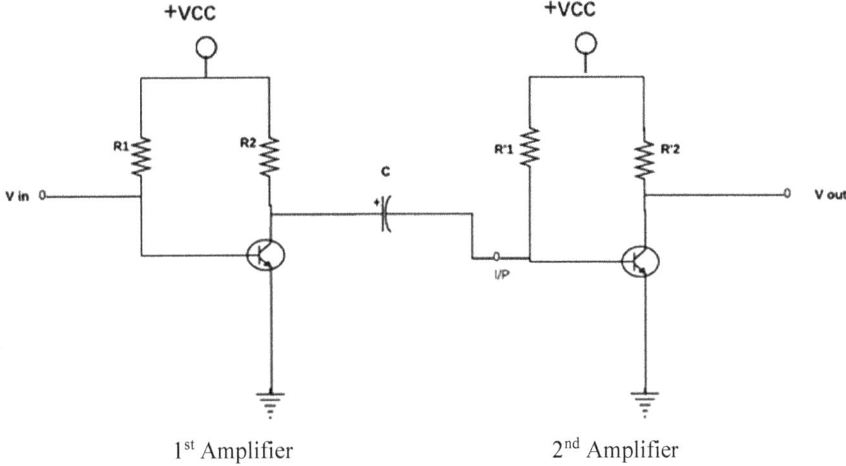

Fig. 7.5 RC coupled amplifier

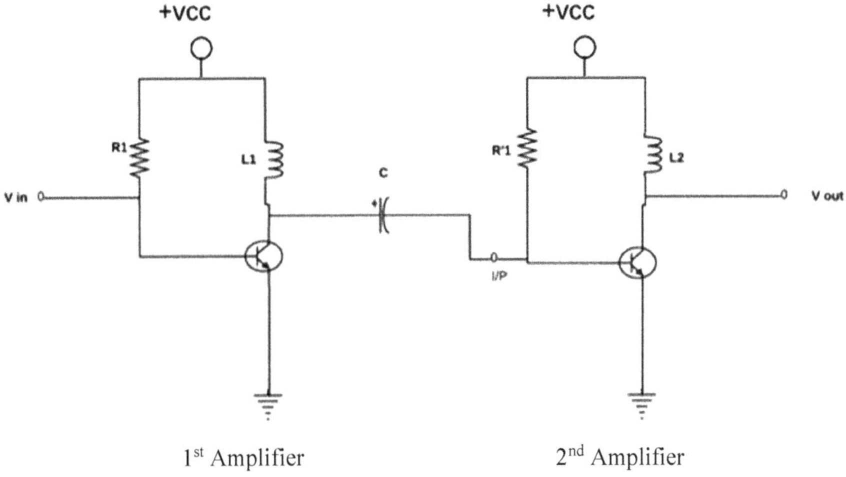

Fig. 7.6 Impedance coupled amplifier

approaches to infinity (∞), i.e., if frequency increase, then X_{L} tends to infinity and each inductor "L" appears V_{in} "open." We can say that inductor passes direct current and block AC one. The amplifier using this coupling scheme is known as "impedance coupling amplifier," and this is useful for high-frequency signal.

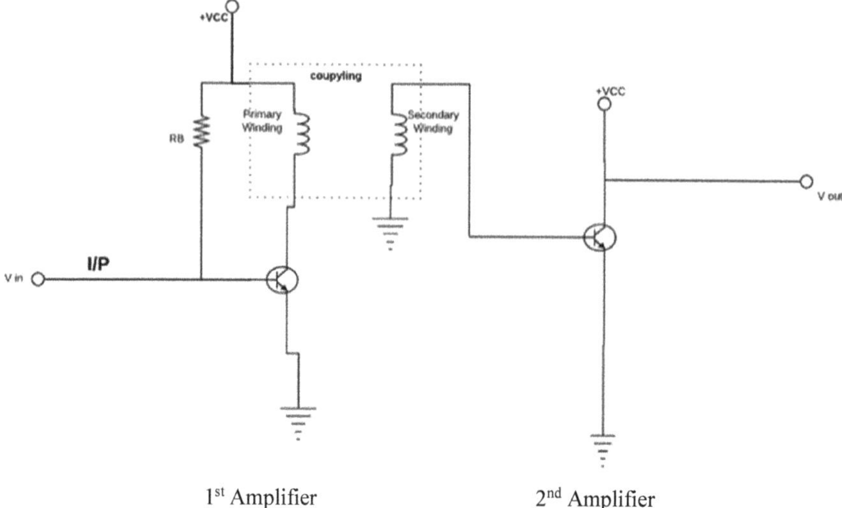

Fig. 7.7 Transformer coupled amplifier

7.4.3 Transformer Coupling

In transfer coupling scheme, the primary winding of the transformer acts as a collector load and the secondary winding conveys the AC output signal directly to the base of the next stage. It can be noted that there is no need of coupling capacitor in the transformer coupling. The amplifiers using this coupling scheme are known as "transformer coupled amplifier" as shown in Fig. 7.7.

7.4.4 Direct Coupling or DC Amplifier Coupling

In this method, the AC output signal is fed directly to the next stage. This type of coupling is used where low-frequency signals are to be amplified. The coupling devices such as capacitors "*C*," inductors "*L*," and transformers cannot be used at low frequencies because their sizes become very large. The amplifiers, using this coupling scheme, are known as "direct coupled amplifier or DC amplifier" as shown in Fig. 7.8.

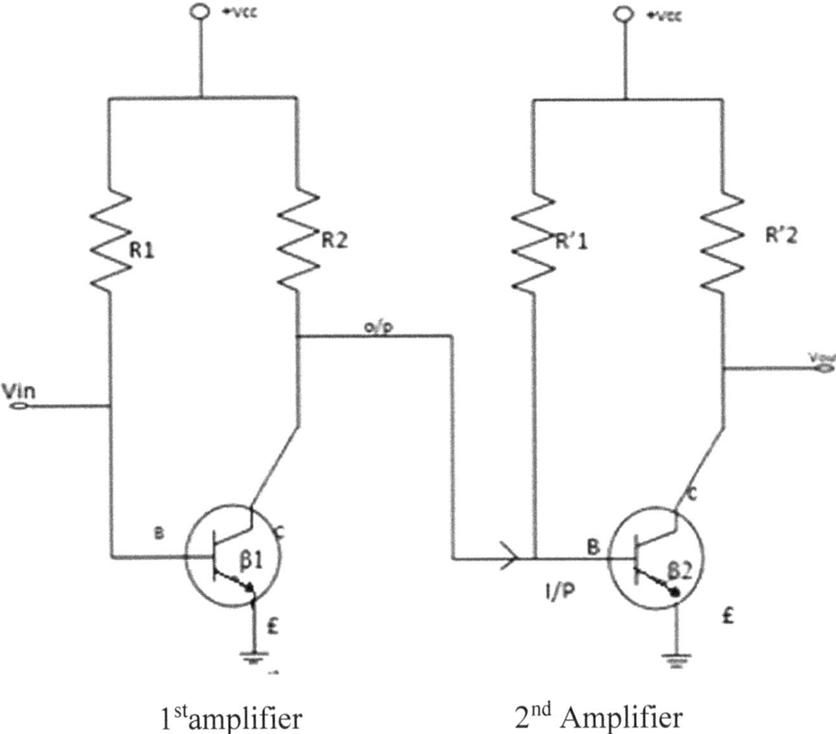

Fig. 7.8 Direct coupled amplifier

7.5 Darlington Amplifier and Its Analysis

7.5.1 Darlington Amplifier

Figure 7.9 illustrates the cascaded emitter followers: NPN Darlington amplifier amplifies an input AC signal. The circuit consists of two cascade emitter followers where the output of first stage transistor (Q_1) goes to input of second transistor (Q_2). This configuration improves amplifier characteristics.

The Darlington amplifier has a high input resistance (R_{in}) and low output resistance (R_{out}), and due to this, the output current across Darlington amplifier increases and the overall current gain also increases. These characteristics make it very useful as a current amplifier, now the current gain can be computed as

$$\text{Current Gain}(\beta) = \frac{\text{Output Current}}{\text{Input Current}} \tag{7.21}$$

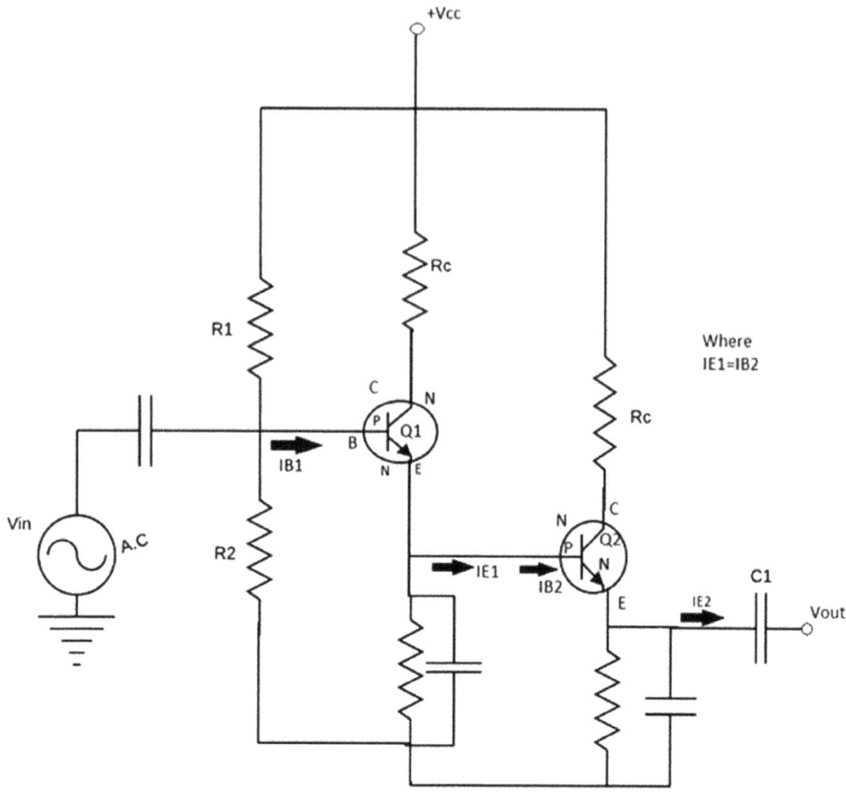

Fig. 7.9 Cascaded emitter followers: NPN Darlington amplifier

These days the semiconductor device manufactures are supplying a pair of identical transistor connected suitably in a single package. It is called Darlington pair.

The two transistors are connected in such a way that the emitter of one transistor connected to the base of the other as shown in Fig. 7.9. The Darlington pairs shown in Fig. 7.10a, b are known as "NPN and PNP Darlington pair," respectively. A Darlington pair is a three-terminal device, namely base (B), emitter (E), and collector (C). It acts like a transistor with an extremely high current gain (β). The Darlington pairs are available for both NPN and PNP transistors.

(i) **Voltage Gain**: The voltage gain of a Darlington amplifier is less than unity, i.e., "1."

$$A_V < 1 \tag{7.22}$$

where A_V represents the voltage gain of Darlington amplifier.

1. Z_{in} of Darlington pair is (β^2 R_E).
2. Current gain of Darlington pair is (β^2).

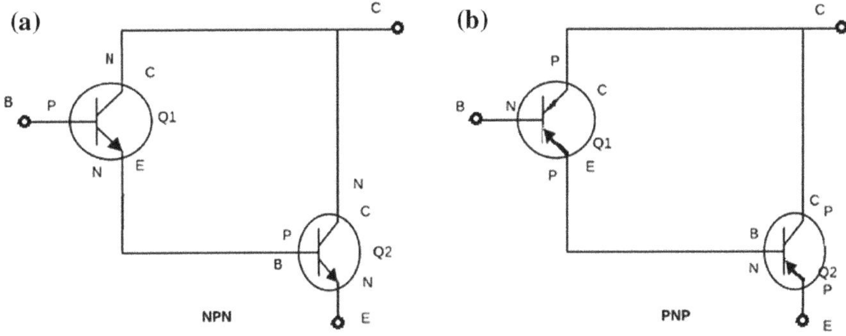

Fig. 7.10 **a** NPN transistor Darlington pair, **b** PNP transistor Darlington pair

3. Voltage in gain of Darlington pair is unity, i.e., (1).

   I_{b1} = Base current of first stage transistor (Q_1).
   I_{e1} = Emitter current of first stage transistor (Q_1).
   β_1 = Current gain of first stage transistor (Q_1).
   I_{b2} = Base current of second transistor (Q_2), its value is I_{e1} of first stage transistor (Q_1), i.e., ($I_{b2} \cong I_{e1}$)
   I_{e2} = Emitter current of second transistor (Q_2).
   β_2 = Current gain of second transistor (Q_2).

(ii) **Current Gain**: We know that the emitter current of the first stage transistor (Q_1) is given as

$$I_{e1} = \beta_1 \times I_{b1} \quad (7.23)$$

and emitter current (I_e) of second transistor (Q_2) is given by

$$I_{e2} = \beta_2 \times I_{b2}. \quad (7.24)$$

As we know that ($I_{b2} = I_{e1}$)
Then,

$$I_{e2} = \beta_2 \times I_{e1}. \quad (7.25)$$

7.5.2 Analysis/Characteristics of a Darlington Amplifier

In Fig. 7.9, I_{b1} shows base current of transistor Q_1, I_{e1} shows emitter current of transistor Q_1, β_1 shows current gain of transistor Q_1, I_{b2} shows base current of transistor Q_2, its value is equal to I_E of Q_1, i.e., ($I_{b2} \cong I_{E1}$), I_{e2} shows emitter current of transistor Q_2, and β_2 show current gain of transistor Q_2.

(i) **Current Gain**: We know that the emitter current of transistor Q_2 is given as

$$I_{e1} = \beta_1 I_{b1} \tag{7.26}$$

and emitter current of transistor Q_2 is given as

$$I_{e2} = \beta_2 I_{b2} = \beta_2 I_{e1}, \quad \text{where,} \quad (I_{b2} = I_{e1}) \tag{7.27}$$

Or $I_{e2} = \beta_2 I_{e1}$

Then,

$$I_{e2} = \beta_2(\beta_1 I_{b1}), \quad \text{using Eq. (7.26)}$$

Therefore, $I_{e2} = \beta^2 I_{b1}$, where $\beta_1 = \beta_2$, and both are constant.

Now, let both the transistors are identical. Then, current gain β_1 and β_2 will be equal. In that case, the emitter current of Q_2 transistor will be

$$I_{e2} = \beta^2 I_{b1} \tag{7.28}$$

Therefore, the overall gain will be

$$A_i = \frac{\text{output}}{\text{input}} = \frac{I_{e2}}{I_{b1}} = > \beta^2 \tag{7.29}$$

The above Eq. (7.29) shows that Darlington amplifier behaves like a single transistor having current gain equal to β^2.

(ii) **Input Resistance**: We know that input resistance of second stage is given by the expression

$$R_{i2} = \beta_2\left(r'_{e1} + R_E\right) \cong \beta_2 R_E$$

Let assuming that $R_E \gg r'_{e1}$

Then,

$$R_{i2} = \beta_2 R_E \tag{7.30}$$

It is the value of resistor R seen by emitter of Q_1 transistor and is given by the expression

$$R'_{i1} = \beta_1\left(r'_{e1} + R_{i2}\right)$$

If $(R_{i2} \gg r'_{e1})$
Then,

$$R'_{i1} = \beta_1 R_{i2} \tag{7.31}$$

As $R_{i2} = \beta_2 R_E$, hence, putting value of R_{i2} in Eq. (7.31), we get
Then,

$$R'_{i1} = \beta_1(\beta_2 R_E) \tag{7.32}$$

For identical transistor, we know that current gains β_1 and β_2 are also equal. Hence, input resistance will be

$$R'_{i1} = \beta^2 R_E \because \beta_1 = \beta_2 = \text{constant} \tag{7.33}$$

and

$$R_{i1} = (R_1 \| R_2) \| \left(R'_{i1}\right)$$

Using Eq. (7.33), we get

$$R_{i1} = (R_1 \| R_2) \| \left(\beta^2 R_E\right), \text{ if } (R_1 \| R_2) \gg \beta^2 R_E$$
$$R_{i1} = (R_1 \| R_2) \tag{7.34}$$

(iii) **Output Resistance**: The output R of the first stage is given by the expression $R_{O1} = r_{e,} + \frac{R_1 \| R_2}{\beta_2}$, and output resistance "R_{O2}" of the second stage will be

$$R_{O2} = r'_{e_2} + \frac{r'_{e1} + \frac{R_1 \| R_2}{\beta_2}}{\beta_1}$$
$$R_{O2} = r'_{e_2} + \frac{r'_{e_1}}{\beta_2} + \frac{R_1 \| R_2}{\beta_1 . \beta_2} \left(\text{if } r'_{e_2} + \frac{r'_{e_1}}{\beta_2} + \frac{(R_1 \| R_2)}{\beta_1 . \beta_2} \text{ are regligible} \right).$$

or,

$$R_{O2} = r'_{e_2} \tag{7.35}$$

(iv) **Voltage Gain**: The voltage gain of Darlington amplifier is given by the expression $A_v = \frac{V_O}{V_{in}}$

The value of output voltage V_O of a Darlington amplifier will be

$$V_O = I_{e_2} . R_E \tag{7.36}$$

and input voltage will be.

$$V_{\text{in}} = I_{\text{e}_1} . r'_{\text{e}_1} + I_{\text{e}_2}\left(r'_{\text{e}_2} + R_{\text{E}}\right) \tag{7.37}$$

The above expression is obtained by applying Kirchhoff's voltage law "(KVL)" to the input circuit of the Darlington amplifier, substituting the value of $I_{\text{e}_1}\left[I_{\text{e}_1} \approx \frac{I_{\text{e}2}}{\beta_2}\right]$ in last equation. We get

$$V_{\text{in}} = \frac{I_{\text{e}2}}{\beta_2} r'_{\text{e}_1} + I_{\text{e}_2}\left(r'_{\text{e}_2} + R_{\text{E}}\right) \tag{7.38}$$

Therefore, voltage gain will be

$$A_{\text{V}} = \frac{I_{\text{e}_2} . R_{\text{E}}}{\frac{I_{\text{e}2}}{\beta_2} . r'_{\text{e}_1} + I_{\text{e}_2}\left(r'_{\text{e}_2} + R_{\text{E}}\right)} \tag{7.39}$$

$$A_{\text{v}} = \frac{R_{\text{E}}}{\frac{r'_{\text{e}1}}{\beta_2} . + \left(r'_{\text{e}_2} + R_{\text{E}}\right)} => \left(\frac{R_{\text{E}}}{R_{\text{E}}}\right), \quad \text{if}\left[R_{\text{E}} \gg r'_{\text{e}_2} \,\&\, \frac{r'_{\text{e}_1}}{\beta_2}\right] \tag{7.40}$$

and,

$$A_{\text{V}} = 1. \tag{7.41}$$

7.5.3 *Analysis of Darlington Amplifier Using h-Parameters*

Let us again consider the circuit of Darlington amplifier with voltage divider bias as shown in Fig. 7.11.

7.5.3.1 Characteristics of a Darlington Amplifier

(i) **Current Gain**: For the second stage of Darlington amplifier, we have input impedance

$$Z_{\text{in}_2} \cong h_{\text{fe}_2} R_{\text{E}} \tag{7.42}$$

and current gain,

$$A_{\text{i}_2} = \frac{\text{o/p}}{\text{i/p}} = \frac{I_0}{I_2} => \frac{I_{\text{e}_2}}{I_{\text{b}_2}} => h_{\text{fe}_2} \tag{7.43}$$

Based upon a good approximation, we cannot apply the equations to the first stage. The obvious reason is to determine which Z_{in_2} compresses with $\frac{1}{h_{\text{fe}_1}}$; we have

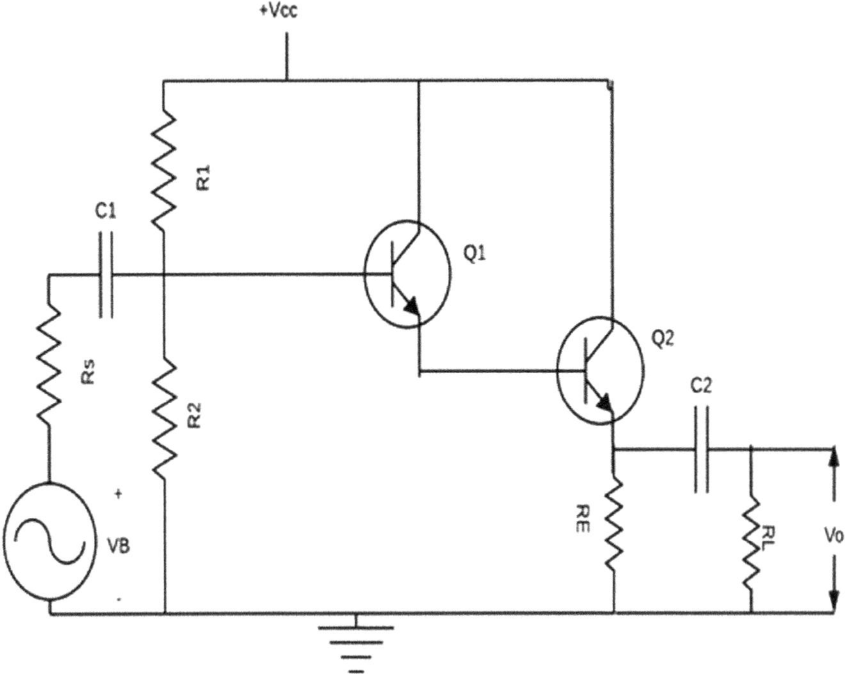

Fig. 7.11 Darlington amplifier circuit in its small signal AC form

observed that 1/hoe could be eliminated in the majority of the cases since the load impedance ($Z_L \ll 1/h_{oe}$).

For the Darlington circuit under consideration, the input impedance Z_{in_2} is close energy in magnitude to $\left(\frac{1}{h_{oe_1}}\right)$ to necessitate considering the effects of h_{oe_1}. We have that for a single-stage grounded emitter T_X amp, with $\frac{1}{h_{oe}}$ taken into consideration.

$$A_i = \frac{h_{fe}}{1 + h_{oe}Z_L} \tag{7.44}$$

Applying the above equation to this situation, we have.

$$Z_L = Z_{in_2} \cong h_{fe_2}R_E \tag{7.45}$$

$$\text{and} \quad A_{i_1} = \frac{I_2}{I_1} = \frac{I_{e_1}}{I_{b_1}} \cong \frac{h_{fe_1}}{1 + h_{oe_1}\left(h_{fe_2}R_E\right)} \tag{7.46}$$

Therefore, the mathematical expression for the current gain is

$$A_i = \frac{I_0}{I_1} => A_{i_1} \times A_{i_2}$$

Using Eqs. (7.43), (7.44), and (7.45), thus

$$A_{\mathrm{i}} = \frac{h_{\mathrm{fe}_1} \times h_{\mathrm{fe}_2}}{1 + h_{\mathrm{oe}_1} \times \left(h_{\mathrm{fe}_2} \times R_{\mathrm{E}}\right)} \tag{7.47}$$

for $h_{\mathrm{fe}_1} = h_{\mathrm{fe}_2} = h_{\mathrm{fe}}$,
and $h_{\mathrm{oe1}} = h_{\mathrm{oe}}$,
The current gain A_{i} will be

$$A_i = \frac{h_{\mathrm{fe}}^2}{1 + h_{\mathrm{oe}} h_{\mathrm{fe}} R_{\mathrm{E}}} \tag{7.48}$$

Let assuming that $h_{\mathrm{oe}} h_{\mathrm{fe}} R_{\mathrm{E}} \leq 0.1$, we can have a very good approximation (within 10%) as under

$$A_{\mathrm{i}} \cong h_{fe}^2 = \beta^2 \quad \text{(using Eq. (7.29))} \tag{7.49}$$

Alternatively, we can obtain the current gain as under,
Emitter current,

$$I_{\mathrm{E2}} = (1 + \beta_2) \times I_{\mathrm{B2}} = (1 + \beta_2) \times I_{\mathrm{E1}} \quad (I_{\mathrm{B2}} = I_{\mathrm{E1}}) \tag{7.50}$$

$$I_{\mathrm{E2}} = (1 + \beta_2) \times I_{\mathrm{E1}} \tag{7.51}$$

$$\text{As,}\ \ I_{\mathrm{B2}} = I_{\mathrm{E1}},\ \ \text{then}\ I_{\mathrm{E1}} = (1 + \beta_2) \times I_{\mathrm{B2}} \tag{7.52}$$

Therefore,

$$I_{\mathrm{E2}} = (1 + \beta_2) \times (1 + \beta_1) \times I_{\mathrm{B1}} \tag{7.53}$$

Hence, current gain, A_i will be,

$$Z_{\mathrm{in1}} = \frac{I_{\mathrm{E2}}}{I_{\mathrm{B1}}} = (1 + \beta_2) \times (1 + \beta_1) \tag{7.54}$$

$$A_i \cong \beta_1 \times \beta_2 \cong \beta^2 \ (\text{if}\ \beta_1 = \beta_2) \tag{7.55}$$

(ii) **Input Impedance**: Because $Z_{\mathrm{in2}} \cong h_{\mathrm{fe2}} R_{\mathrm{E}}$ is the emitter resistance of the first stage. Therefore, the input impedance to the first stage will be

$$Z_{\mathrm{in1}} \cong h_{\mathrm{fe1}} \left(Z_{\mathrm{in2}} \| \frac{1}{h_{\mathrm{oe1}}} \right) \tag{7.56}$$

Since $Z_{\text{in2}} \cong h_{\text{fe2}} R_{\text{E}}$, and $\frac{1}{h_{oe1}}$ would appear in parallel in the small signal equivalent circuit. Therefore, $Z_{\text{in1}} = h_{\text{fe1}}\left(h_{\text{fe2}} R_{\text{E}} \| \frac{1}{h_{\text{oe1}}}\right)$

$$Z_{\text{in1}} = \left[\frac{h_{\text{fe1}} h_{\text{fe2}} \frac{R_{\text{E}}}{h_{\text{oe1}}}}{h_{\text{fe2}} R_{\text{E}} + \frac{1}{h_{\text{oe1}}}}\right] \tag{7.57}$$

Or

$$Z_{\text{in1}} = \frac{h_{\text{fe1}} h_{\text{fe2}} R_{\text{E}}}{h_{\text{oe1}} h_{\text{fe2}} R_{\text{E}} + 1} \tag{7.58}$$

Now, if $h_{\text{fe1}} = h_{\text{fe2}}$ and $h_{\text{oe1}} = h_{\text{oe}}$, then the input impedance will be

$$Z_{\text{in1}} = \frac{h_{\text{fe}}^2 R_{\text{E}}}{1 + h_{\text{oe}} h_{\text{fe}} R_{\text{E}}} \tag{7.59}$$

Let assuming that $h_{\text{oe}} h_{\text{fe}} R_{\text{E}} \leq 0.1$, then

$$Z_{\text{in1}} \cong h_{\text{fe}}^2 R_{\text{E}} = \beta^2 R_{\text{E}}, \quad \text{(using Eq. (7.49))} \tag{7.60}$$

Alternatively, we can obtain the input impedance as under,
Input impedance of second stage is given by,

$$Z_{\text{in2}} = \beta_2 \left(R_{\text{E}} + r'_{\text{e2}}\right) \tag{7.61}$$

where r'_{e2} is the AC emitter resistance of the second transistor, if $R_{\text{E}} \gg r'_{\text{e2}}$
Then, we have,

$$Z_{\text{in2}} = \beta_2 R_{\text{E}} \tag{7.62}$$

Z_{in2} is also the output impedance while looking into the emitter of the first transistor. Hence, the input impedance looking into the base of the first transistor will be,

$$Z_{\text{in1}} = \beta_1 \left(r'_{\text{e1}} + Z_{\text{in2}}\right) => \beta_1 \left(r'_{\text{e1}} + \beta_2 R_{\text{E}}\right) \tag{7.63}$$

or,

$$Z_{\text{in1}} \cong \beta_1 \beta_2 R_{\text{E}} \cong \beta^2 R_{\text{E}} \quad \text{if, } \beta_1 = \beta_2 = \beta \tag{7.64}$$

This input impedance is very high because of the product of two gains, and therefore input impedance, $Z_{\text{in}} = R_1 \| R_2 \| Z_{\text{in1}} \cong R_1 \| R_2$ where $(Z_{\text{in1}} \gg R_1 \| R_2)$

Also, if there is a load resistance (R_L) coupled to the transistor of the second transistor, then we have,

$$Z_{\mathrm{in1}} = \beta^2 (R_{\mathrm{E}} \| R_{\mathrm{L}})$$
$$Z_{\mathrm{in1}} \cong \beta^2 R_{\mathrm{E}} \tag{7.65}$$

(iii) **Output Impedance**: The output impedance can be determined directly from the emitter equivalent circuits as under,

$$Z_{\mathrm{o1}} = \frac{R_{\mathrm{S}} + h_{\mathrm{ie1}}}{h_{\mathrm{fe1}}} \text{ and, } \quad Z_{\mathrm{o2}} = \frac{\left[Z_{\mathrm{o1}} \| \frac{1}{h_{\mathrm{oe1}}} + h_{\mathrm{ie2}} \right]}{h_{\mathrm{fe2}}} \tag{7.66}$$

Alternatively, we can be determined the output impedance as under, the AC Thevenin's impedance at the input will be

$$r_{\mathrm{th}} = R_{\mathrm{S}} \| R_1 \| R_2$$

Since, by shorting all voltage source R_1 and R_{S} come in parallel with R_2. Thus, output impedance of the first stage is,

$$Z_{\mathrm{o1}} = r'_{\mathrm{e1}} + \frac{r_{\mathrm{th}}}{\beta_1}$$

Output impedance of second stage is,

$$Z_{\mathrm{o2}} = r'_{\mathrm{e2}} + \frac{Z_{\mathrm{o1}}}{\beta} => r'_{\mathrm{e2}} + \left[\frac{r'_{\mathrm{e1}} + \frac{r_{\mathrm{th}}}{\beta_1}}{\beta_2} \right] \tag{7.67}$$

Z_{o} is quite smaller than Z_{o1}, i.e., ($Z_{\mathrm{o}} \ll Z_{\mathrm{o1}}$), i.e., Z_{o} is lowered. Due to this, Darlington amplifier can be used to isolate high impedance source from low impedance load. If high impedance is directly connected to a low impedance load, most of the signal voltage will be dropped across the high impedance of the source and the remaining source signal available may not be able to drive the load.

(iv) **Voltage Gain**: Applying "KVL" to the circuit shown in Fig. 7.12, we have $V_{\mathrm{O}} = \left(V_{\mathrm{i}} - V_{\mathrm{be_1}} - V_{\mathrm{be_2}} \right)$. This means that the output potential is the input potential less than the base-to-emitter potential of each transistor. This clearly indicates that ($V_{\mathrm{O}} < V_{\mathrm{in}}$), and voltage gain is closer in magnitude to one than to zero.

Further, on an approximate basis, it is given as,

$$A_{\mathrm{v}} \cong \left[\frac{1}{1 + \frac{h_{\mathrm{ie_2}}}{h_{\mathrm{fe_2}} \times R_{\mathrm{E}}}} \right] \tag{7.68}$$

Also, which is on an approximate basis, the v-gain can be determined from the following relation.

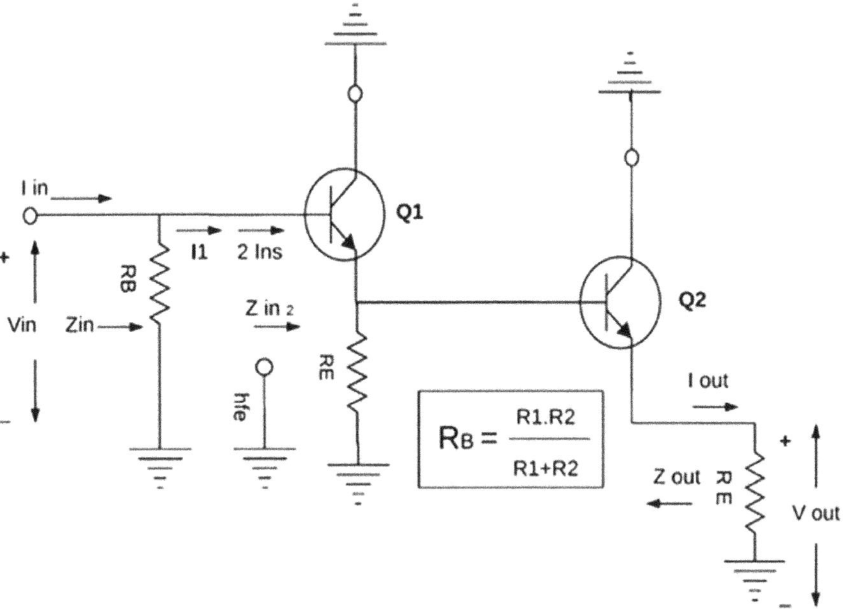

Fig. 7.12 Small signal AC for Darlington amplifier

$$A_V \cong \frac{R_E}{r'_e + R_E} => \frac{1}{1 + \frac{r'_e}{R_E}}, \quad \text{if} \left(\frac{r'_e}{R_E} \ll 1 \right) \tag{7.69}$$

Then,

$$A_V \cong 1 \tag{7.70}$$

Which is slightly less than "unity," as in case of "emitter follower."

Question 7.3 Determine the gain of the given values

1. If voltage gain = 40 dB.
2. If power gain = 44.772 dB.

Solution

(1) Voltage Gain|dB $= 20 \times \log_{10}\left(\frac{V_0}{V_{in}}\right)$

$$40|\text{dB} = 20 \times \log_{10}(\text{Gain}|_{\text{without dB}})$$

$$10^2 = \text{Gain}|_{\text{without dB}}$$

$$\text{and,} \quad \text{Gain}|_{\text{without dB}} = 100.$$

(2) $\text{Power Gain}|_{\text{without dB}} = 10\log_{10}\left(\frac{P_0}{P_{\text{in}}}\right)$

$$\begin{aligned}44.772|\text{dB} &= 10 \times \log_{10}(\text{Gain}|_{\text{without dB}})\\ \text{Gain}|_{\text{without dB}} &= 10^{4.4772}\\ \text{Gain}|_{\text{without dB}} &= 30005.44002\end{aligned}$$

Question 7.4 A multistage amplifier consists of four stages each of which has a voltage gain of 40

1. What is the overall gain of the amplifier?
2. If one stage of the four stages having negative feedback, then what will be the resultant gain.
3. If fifth stage gives negative feedback of 10 dB, find out the overall gain.

Solution Total number of stages $n = 4$, each gain is 40 dB,
So, Voltage Gain $A_1 = A_2 = A_3 = A_4 = 40$.

(1) $$\begin{aligned}\text{Voltage Gain}|\text{in dB} &= 20\log_{10}(40)\\ &= 32.04\end{aligned}$$

$$\begin{aligned}\text{Overall gain} &= A_1 \times A_2 \times A_3 \times A_4\\ &= (32.04 + 32.04 + 32.04 + 32.04)\\ \text{Overall gain} &= 128.16\,\text{dB}.\end{aligned}$$

(2) If one stage of four stages having negative feedback, then

$$\begin{aligned}\text{Overall gain will be} &= (32.04 + 32.04 + 32.04 - 32.04)\\ &= 64.08\,\text{dB}.\end{aligned}$$

(3) If fifth stage gives negative gain of 10 dB, then

$$\begin{aligned}\text{Overall gain} &= (32.04 + 32.04 + 32.04 + 32.04 - 10)\text{dB}\\ &= 128.16 - 10\\ &= 118.16\,\text{dB}.\end{aligned}$$

7.6 Bootstrapping Amplifier

As a matter of fact, the maximum R_{in} of a Darlington circuit is limited to $\frac{1}{h_{\text{ob}}} \cong 2\,\text{m}\Omega$ as $\frac{1}{h_{\text{eo}}}$ is the resistance between the base and the collector.

However, the input resistance can be largely increased by bootstrapping the Darlington circuit through the addition of capacitor "C" between the first collector terminal C_1 and the second emitter terminal E_2 as shown in Fig. 7.13, and it may be noted that the resistor R_C is essential because of the absence of R_C, resistor R_E will be shorted to ground. If the input signal changes by V_{in}, then E_2 changes by $(A_v V_{in})$. Assuming that reactance of C' is negligible and the collector changes by the same amount. This means that $\frac{1}{h_{ob}}$ is now effectively increased to $\frac{1}{h_{ob}(1-A_v)} \cong 400\,\text{M}\Omega$ for a voltage gain of 0.995.

We can obtain an expression for the input resistance of the bootstrapped Darlington circuit by using the equivalent circuit shown in Fig. 7.14. The effective resistance R_{eff} between terminal E_2 and ground will be $R_E \| R_C$.

If $(h_{oe} R_{eff} \le 0.1)$, then the transistor Q_2 can be represented by approximate hybrid model, but we must use exact hybrid model for transistor Q_2 as shown in Fig. 7.14 because $\frac{1}{h_{oe1}} \gg h_{ie2}$; therefore, h_{oe1} can be omitted from Fig. 7.13.

Thus, solving for $\frac{V_{in}}{I_{b1}}$, we get,

$$R_{in} = h_{fe1} \times h_{fe2} \times R_{eff} \tag{7.71}$$

Equation (7.71) shows that input resistance of the bootstrapped Darlington circuit is essentially equal to the product of the short circuit current gain and the effective

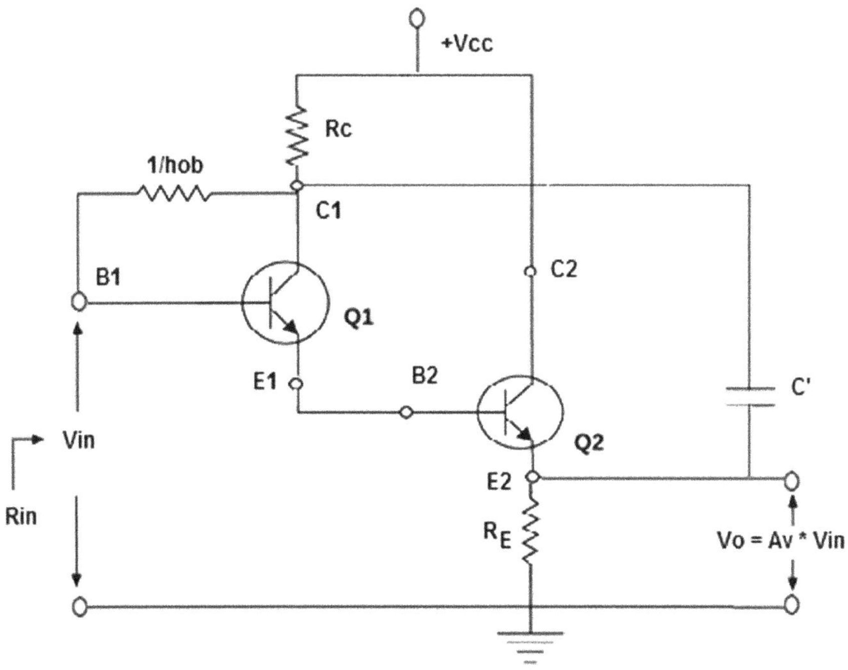

Fig. 7.13 Bootstrapped Darlington amplifier

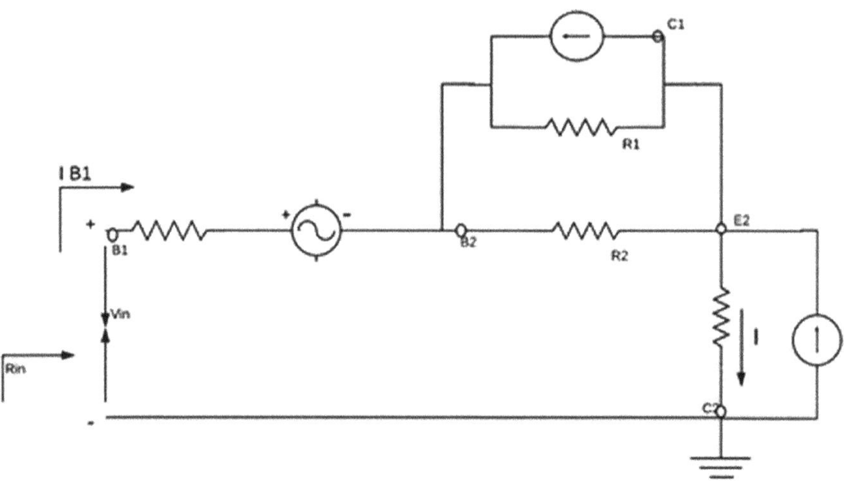

Fig. 7.14 Equivalent circuit for a bootstrapped Darlington amplifier

emitter resistance. If the transistor with current gain of order of magnitude of 100 is used and effective resistance as 5 KΩ, then the R_{in} obtained will be of the order of 50 MΩ.

7.7 Tuned and Double-Tuned Voltage Amplifiers

To amplify the selective range of frequencies, the resistive load, R_C, is replaced by a tuned circuit. The tuned circuit is capable of amplifying a signal over a narrowband of frequencies centered at f_r. The amplifier with such a tuned circuit as a load is known as tuned amplifier.

Figure 7.15 shows the tuned parallel LC circuit which resonates at a particular frequency. The resonance frequency and impedance of tuned circuit are given as

$$f_r = \frac{1}{2\pi\sqrt{LC}} \tag{7.72}$$

$$Z_r = \frac{L}{CR} \tag{7.73}$$

The response of tuned amplifiers is maximum at resonant frequency, and it falls sharply for frequencies below and above the resonant frequency, as shown in Fig. 7.16.

As shown in Fig. 7.16, 3 dB bandwidth is denoted as B and 30 dB bandwidth is denoted as S. The ratio of the 30 dB bandwidth (S) to the 3 dB bandwidth (B) is known as skirt selectivity.

Fig. 7.15 Tuned circuit

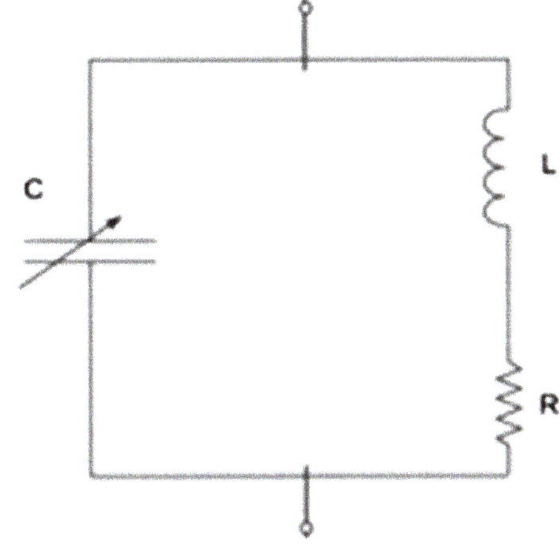

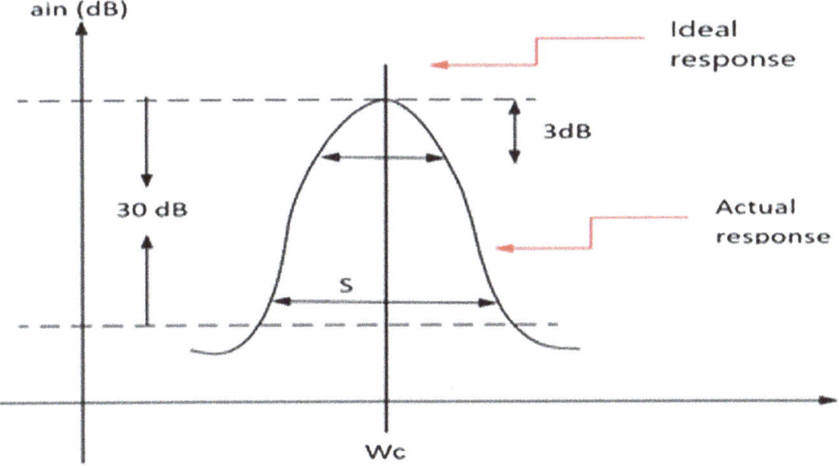

Fig. 7.16 Frequency response of a tuned amplifier

7.7.1 Classification of Tuned Amplifier

The cascaded stages of multistage tuned amplifiers can be categorized as given below

1. Single-tuned amplifier.
2. Double-tuned amplifier.
3. Stagger-tuned amplifier.

7.7.1.1 Single-Tuned Amplifier

A common emitter amplifier can be converted into a single-tuned amplifier by including a parallel tuned circuit as shown in Fig. 7.17. The biasing components are not shown for simplicity.

Before going to study the analysis of this amplifier, we see the several practical assumptions to simplify the analysis.

Assumptions:

1. $R_L \ll R_C$.
2. $r_{bb'} = 0$.

With these assumptions, the simplified equivalent circuit for a single-tuned amplifier is as shown in Fig. 7.18.

Where

$$C_{eq} = C' + C_{b'e} + (1 + g_m R_L) C_{b'c}. \tag{7.74}$$

C' is external capacitance used to tune the circuit.
$(1 + g_m R_L) C_{b'c}$ is the miller capacitance.
r_s represents the losses in the coil.

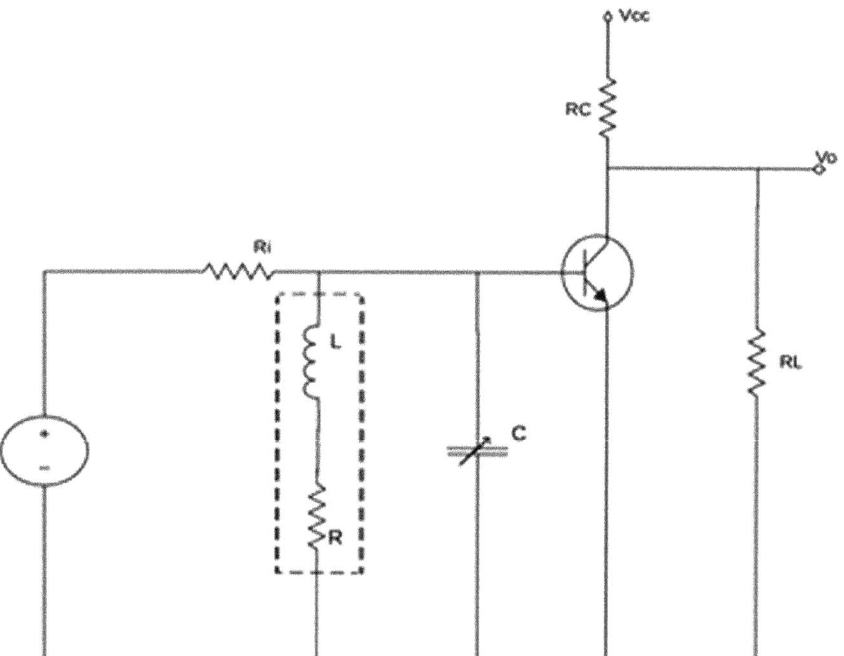

Fig. 7.17 Single-tuned transistor amplifier

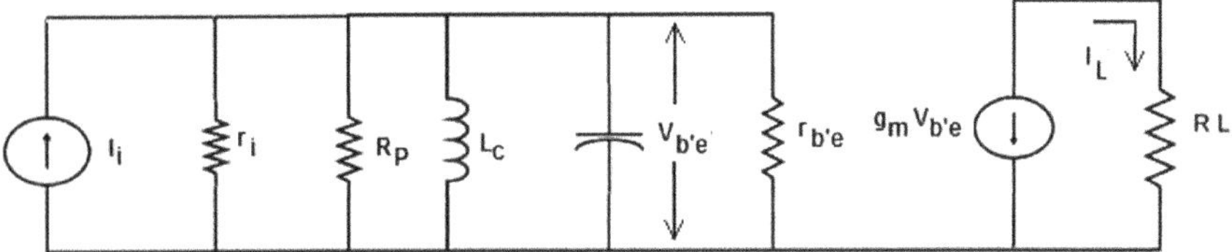

Fig. 7.18 Equivalent circuit of a single-tuned amplifier

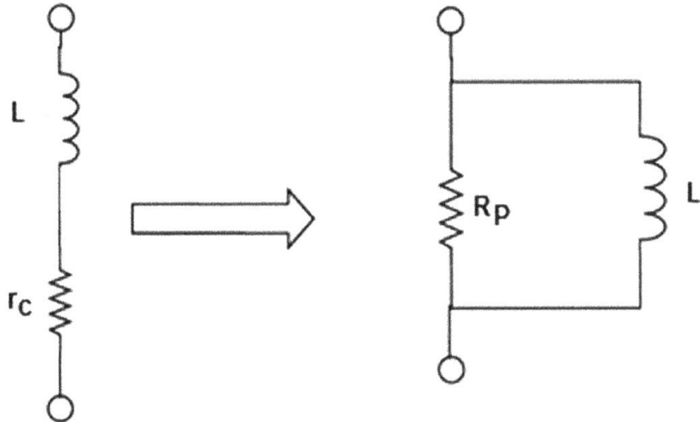

Fig. 7.19 Equivalent circuits

The series RL circuit in Fig. 7.19 is replaced by the equivalent RL circuit in Fig. 7.18, assuming coil losses are low over the frequency band of interest, i.e., the coil Q high.

$$Q_C = \frac{wL}{r_C} \gg 1 \tag{7.75}$$

The conditions for equivalence are most easily established by equating the admittance of the two circuits which are shown in Fig. 7.19.

$$Y_1 = \frac{1}{r_C + jwL} => \frac{r_C - jwL}{r_C^2 + w^2L^2} \tag{7.76}$$

$$Y_1 = \frac{r_C}{r_C^2 + w^2L^2} - \frac{jwL}{r_C^2 + w^2L^2} \tag{7.77}$$

From Eq. (7.77), $wL \gg r_C$, then

$$Y_1 = \frac{r_C}{w^2L^2} - \frac{j}{wL} \tag{7.78}$$

$$Y_1 = \frac{r_C}{w^2 L^2} + \frac{1}{jwL} \tag{7.79}$$

$$Y_2 = \frac{1}{R_P} + \frac{1}{jwL} \tag{7.80}$$

Therefore, equating Y_1 and Y_2, we get

$$\frac{r_C}{w^2 L^2} + \frac{1}{jwL} = \frac{1}{R_P} + \frac{1}{jwL} \tag{7.81}$$

$$\frac{1}{R_P} = \frac{r_C}{w^2 L^2} \tag{7.82}$$

Using Eq. (7.82), we get

$$\frac{1}{R_P} = \frac{r_C^2}{r_C w^2 L^2} \tag{7.83}$$

$$\frac{1}{R_P} = \frac{1}{r_C Q_C^2} \tag{7.84}$$

$$R_P = r_C Q_C^2 => wLQ_C \tag{7.85}$$

Looking at Fig. 7.18, we get

$$R = r_i \| R_P \| r_{b'e} \tag{7.86}$$

Then, the current gain of the amplifier will be

$$A_i = \frac{-g_m R}{1 + j(wRC - R/wL)} \tag{7.87}$$

$$A_i = \frac{-g_m R}{1 + jw_0 RC\left(\frac{w}{w_0} - \frac{w_0}{w}\right)} \tag{7.88}$$

where $w_0^2 = \frac{1}{LC}$

We define the Q of the tuned circuit at the resonant frequency w_0 to be

$$Q_i = \frac{R}{w_0 L} => w_0 RC \tag{7.89}$$

$$A_i = \frac{-g_m R}{1 + jQ_i\left(\frac{w}{w_0} - \frac{w_0}{w}\right)} \tag{7.90}$$

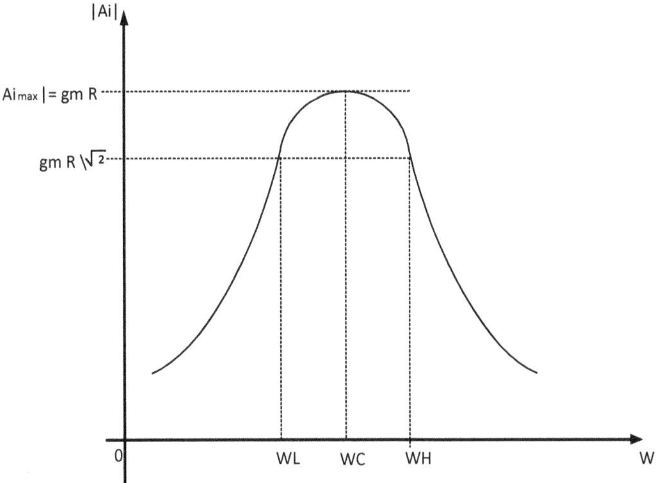

Fig. 7.20 Gain versus frequency for single-tuned amplifier

At $w = w_0$, gain is maximum and it is given as

$$A_{\mathrm{i(maxi)}} = -g_{\mathrm{m}} R \tag{7.91}$$

Figure 7.20 shows the gain vs frequency plot for single-tuned amplifier. It shows the variation of the magnitude of the gain as a function of frequency.

At 3 dB frequency,

$$|A_{\mathrm{i}}| = \frac{g_{\mathrm{m}} R}{\sqrt{2}} \tag{7.92}$$

We can write it as

$$1 + jQ_{\mathrm{i}}\left(\frac{w}{w_0} - \frac{w_0}{w}\right) = \sqrt{2} \tag{7.93}$$

$$1 + Q_{\mathrm{i}}^2\left(\frac{w}{w_0} - \frac{w_0}{w}\right)^2 = 2 \tag{7.94}$$

This equation is quadratic in w^2 and has two positive solutions, w_{H} and w_{L}. Now, we get 3 dB bandwidth as given below

$$\mathrm{BW} = f_{\mathrm{H}} - f_{\mathrm{L}} \Rightarrow \left(\frac{w_0}{2\pi Q_i}\right) = \frac{1}{2\pi RC} \tag{7.95}$$

$$\mathrm{BW} = \frac{1}{2\pi RC} \tag{7.96}$$

7.7.1.2 Double-Tuned Amplifier

A double-tuned amplifier is a tuned amplifier with transformer coupling between the amplifier stages in which the inductances of both the primary and secondary windings are tuned separately with a capacitor across each. The scheme results in a wider bandwidth and steeper skirts than a single-tuned circuit would achieve.

There is a critical value of transformer coupling coefficient at which the frequency response of the amplifier is maximally flat in the pass band and the gain is maximum at the resonant frequency. Designs frequently use a coupling greater than this (over-coupling) in order to achieve an even wider bandwidth at the expense of a small loss of gain in the center of the pass band.

Cascading multiple stages of double-tuned amplifiers results in a reduction in the bandwidth of the overall amplifier. Two stages of double-tuned amplifier have 80% of the bandwidth of a single stage. An alternative to double tuning that avoids this loss of bandwidth is staggered tuning. Stagger-tuned amplifiers can be designed to a prescribed bandwidth that is greater than the bandwidth of any single stage. However, staggered tuning requires more stages and has lower gain than double tuning.

In Fig. 7.21, the circuitry consists of two stages of amplifier in common emitter topology. The bias resistors all serve their usual functions. The input of the first stage is coupled in the conventional way with a series capacitor to avoid affecting the bias. However, the collector load consists of a transformer which serves as the inter-stage coupling instead of capacitors. The windings of the transformer have inductance. Capacitors placed across the transformer windings form resonant circuits which provide the tuning of the amplifier.

A further detail that may be seen in this kind of amplifier is the presence of taps on the transformer windings. These are used for the input and output connections of the transformer rather than the top of the windings. This is done for impedance

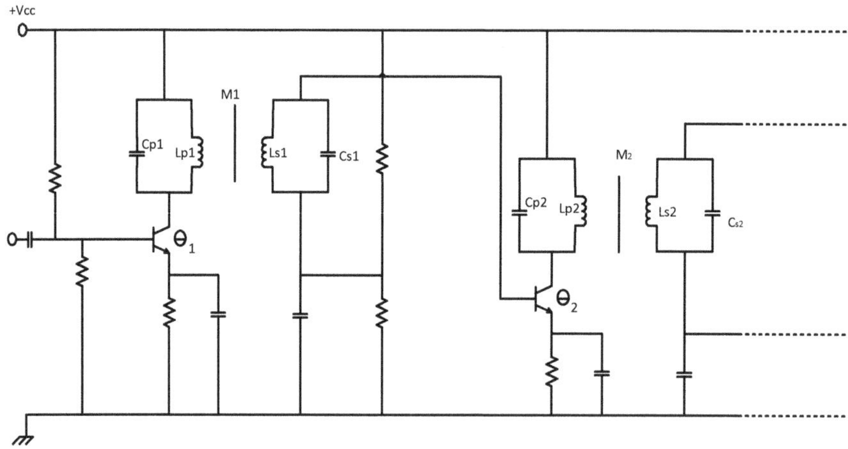

Fig. 7.21 Typical two-stage double-tuned amplifier

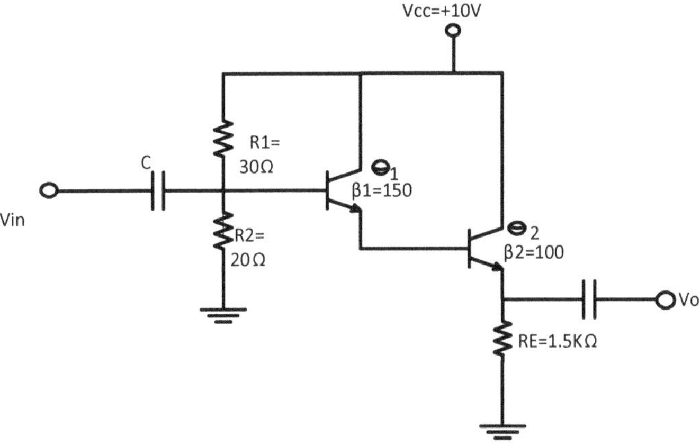

Fig. 7.22 Typical two-stage double-tuned amplifier

matching purposes, and bipolar junction transistor amplifiers (the kind shown in the circuit) have a quite high output impedance and quite low input impedance. This problem can be avoided by using MOSFETs which have very high input impedance.

The capacitors connected between the bottom of the transformer secondary windings and ground do not form part of the tuning. Rather, their purpose is to decouple the transistor bias resistors from the AC circuit.

Question 7.5 Figure 7.22 shows the circuit diagram of a Darlington amplifier, Compute:

(1) The overall gain.
(2) The AC emitter diode resistance for each transistor.
(3) Total input resistance.
(4) Overall voltage gain.

Solution Given that,

$$V_{CC} = 10,\ R_1 = 30\,\text{K}\Omega,\ R_2 = 20\,\text{K}\Omega,\ R_E = 1.5\,\text{K}\Omega,\ \beta_1 = 150,\ \text{and}\ \beta_2 = 100.$$

(1) We know that overall current gain is given as,

$$A_i = \beta_1 \times \beta_2$$
$$A_i = 150 \times 100 \Rightarrow 15{,}000.$$

(2) Voltage drop across R_L will be,

$$V_{R2} = V_{cc} \times \frac{R_2}{R_1 + R_2} \Rightarrow 10 \times \frac{20}{30 + 20} \Rightarrow 4\,\text{V}.$$

And the voltage at the base of Q_2 (or emitter of Q_1) will be

$$V_{B2} = V_{R2} - V_{BE1} \Rightarrow (4 - 0.7) \Rightarrow 3.3\,\text{V}.$$

Similarly, voltage at the emitter of Q_2 will be,

$$V_{E2} = V_{B2} - V_{BE2} \Rightarrow 3.3 - 0.7 \Rightarrow 2.6\,\text{V}.$$

And the value of emitter current of Q_2 will be,

$$I_{E2} = \frac{V_{E2}}{R_E} \Rightarrow \frac{2.6}{1.5 \times 10^3} \Rightarrow 1.73\,\text{mA}.$$

Hence, AC emitter diode resistance of transistor Q_2 will be,

$$r'_{e2} = \frac{25}{I_{E2}(\text{mA})} \Rightarrow \frac{25}{1.73} \Rightarrow 14.5\,\Omega.$$

Now, base current of Q_2 will be,

$$I_{b2} = \frac{I_{E2}}{\beta_2} \Rightarrow \frac{1.73}{100} \Rightarrow 0.0173\,\text{mA}.$$

And the emitter current of Q_1 will be,

$$I_{E1} = I_{b2} = 0.0173\,\text{mA}.$$

Hence, AC emitter diode resistance of Q_1 will be,

$$r'_{e1} = \frac{25}{I_{E1}\text{mA}} \Rightarrow \frac{25}{0.0173} \Rightarrow 1445\,\Omega.$$

(3) We know that total input resistance R_{in} (i.e., input resistance of the amplifier stage will be given by),

$$R_{i1} = R_1 \| R_2 \Rightarrow 30\|20 \Rightarrow 12\,\text{K}\Omega.$$

(4) We also know that overall voltage gain is expressed as,

$$A_v = \frac{R_E}{\frac{r'_{e1}}{\beta_2} + \left(r'_{e2} + R_E\right)} \Rightarrow \frac{1.5 \times 10^3}{\frac{1445}{100} + \left[14.4 + (1.5 \times 10^3)\right]}$$
$$A_v = 0.98.$$

Summary

- **This chapter included introduction of multistage amplifier.**
- **Applications of multistage amplifier and calculated overall gain.**

- **Frequency response of an amplifier.**
- **Define cascade amplifier.**
- **Coupling techniques like resistance, direct, impedance, and transfer coupling.**
- **Discuss Darlington amplifier and its analysis.**
- **Characteristics of a Darlington amplifier.**
- **Define Bootstrapping amplifier.**
- **Tuned and double-tuned voltage amplifiers.**

Numerical Problems

Question 1: What is coupling, draw and explain direct coupling technique.
Question 2: Explain $n = 4$ multistage amplifiers, where "n" is number of stages.
Question 3: For an amplifier midband gain is 50 and lower cut-off frequency is 500 Hz, compute the gain of amplifier at 20 Hz.
Question 4: In the circuit given in Fig. 7.23, has the following components $R_1 =$ 21 K, $R_2 = 12$ K, $R_C = 11$ K, $R_E = 5.5$ K, $R_S = 1.2$ K, and $R_L = 10$ K. The transistor parameters are $h_{fb}= -0.99$, $h_{ob} = 0.5$ µA/V, and $h_{ib} = 20\ \Omega$. Compute R_i, R_i', R_0, R_0', A_i, A_v, and A_{VS}.
Question 5: Explain and draw cascade amplifier.
Question 6: Explain double-tuned voltage amplifier.
Question 7: A transistor amplifier shown in Fig. 7.24 uses a transistor whose h-parameters are as follows $h_{ie} = 1.2$ k, $h_{fe} = 75$, $h_{re} = 2.4 \times 10^{-4}$, $h_{oe} = 25 \times 10^{-6}$ A/V. compute I_0/I_i, A_V, A_{VS}, R_0', and R_i'.
Question 8: Determine the gain of the given values, if voltage gain is 10 dB, and power gain is 20 dB.
Question 9: What is Darlington amplifier, drive and discuss its analysis also.
Question 10: Explain and analyze bootstrapping amplifier.

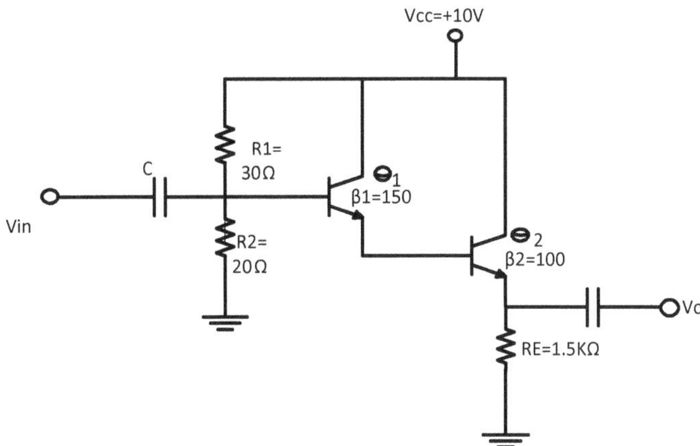

Fig. 7.23 RC Network

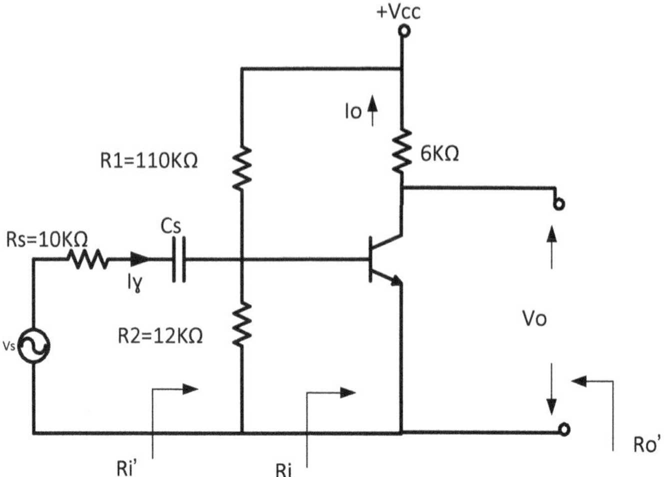

Fig. 7.24 Transistor amplifier

Question 11: Draw the two-stage RC coupled *CE-CE* cascade amplifier.
Question 12: In the given Fig. 7.25, a transistor with $h_{ie} = 1.2\ \text{k}\Omega$, $h_{fe} = 75$, $h_{re} = 200 \times 10^{-4}$, $h_{oe} = 20\ \mu\text{A/V}$ is connected in CE configuration given below. Compute I_O/I_i, I_O/I_S, A_V, A_{VS}, R_O', and R_i'.

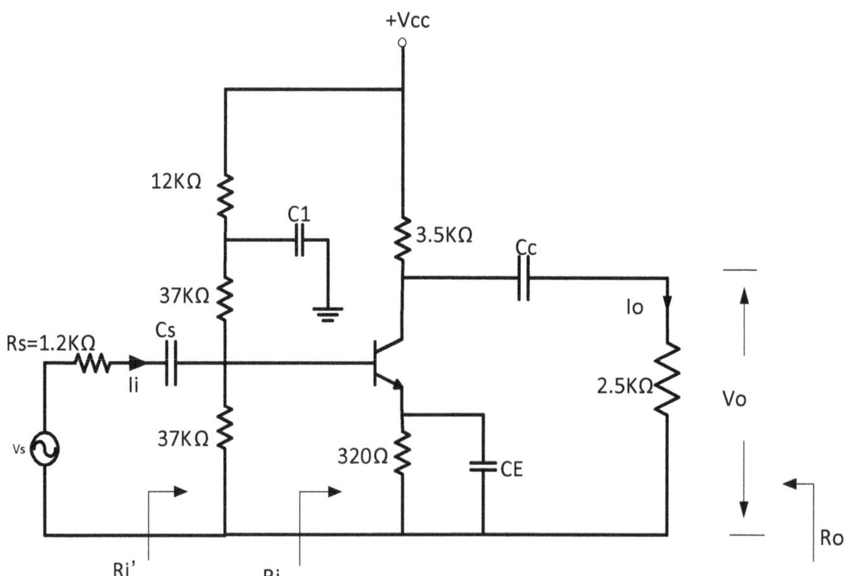

Fig. 7.25 RC Transistor network

www.ingramcontent.com/pod-product-compliance
Ingram Content Group UK Ltd.
Pitfield, Milton Keynes, MK11 3LW, UK
UKHW021834270726
14058UKWH00001B/137

* 9 7 8 9 8 1 1 5 0 2 6 9 9 *